Remarkable Plants
That Shape Our World

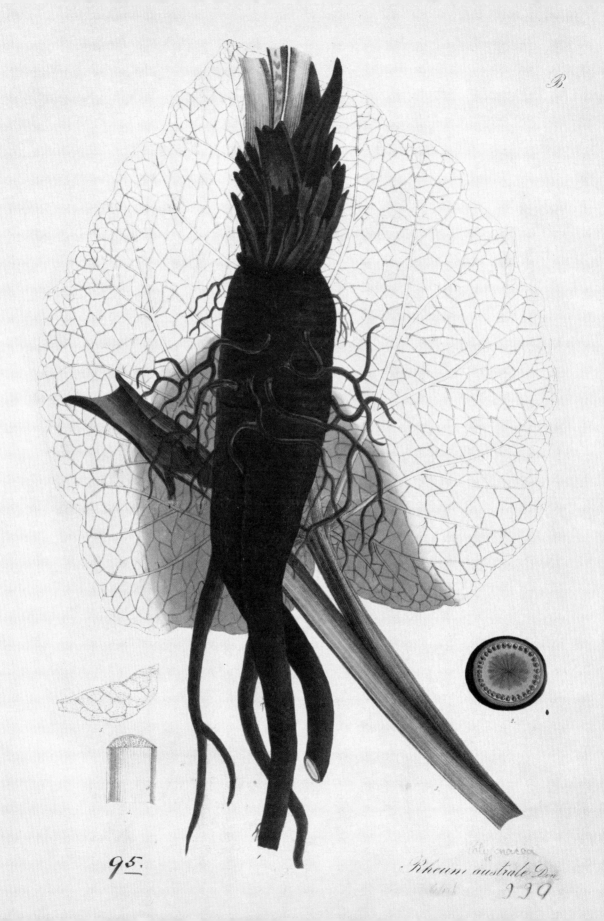

95.

Rheum australe Don.

999

HELEN & WILLIAM BYNUM

Remarkable Plants
That Shape Our World

The University of Chicago Press
Chicago

Contents

Heal and Harm 82
Getting the Balance Right

Technology and Power 112
The Material World

Cash Crops
Making It Pay

Landscape
Plant Aesthetics on the Grand Scale

Revered and Adored 180
From the Sacred to the Exquisite

Wonders of Nature 214
The Extraordinary Plant World

Introduction
Utility and Beauty

Remarkable Plants That Shape Our World is a celebration of the utility, beauty, diversity and sheer wonder of the plant world gracing our planet. For millennia we relied on plants for much of our food, shelter, clothing, transport and medicine. Our roaming ancestors took what was on offer from the wild, like other foraging animals, but when we entered into synergy with certain plants after the last ice age, their domestication at our hands and our novel settled habits began a new era. Although today we enjoy the convenience of the products of the modern petrochemical industry, our need for plants remains keen. They are the basis of all food chains and our great ingenuity hasn't changed this. And in an increasingly urbanized world the capacity of plants to act as green lungs for cities and solace for the soul of the city-dweller is paramount. The areas of untouched wildness are few and far between and therefore more precious than ever.

Such a latter-day fusion of utility and beauty has deep roots. It reflects the ways humans have long engaged with plants, from appreciating their purely practical benefits to responding to their stimulation of the senses and swaying of the emotions. Plants have been instrumental in the development of cultures and even empires. They have been worshipped and deified; their forms, colours and scents have inspired a desire to own and grow certain ones amounting to an obsession – think of tulips, roses and orchids. Other plants have elevated food beyond mere sustenance to the heady delights of rarefied taste. Essential oils and resins have perfumed our bodies and some chemical constituents have proved remarkably efficacious medicinally or have taken our internal neural chemistry in exotic, perhaps frightening directions.

The green kingdom itself is marvellous. Plants have the ability to capture the energy of the sun and use it to power a chemical reaction in their cells which combines water and carbon dioxide to produce sugars while releasing oxygen. This is the process we call photosynthesis. Plants can perform this alchemy because at some time in their ancestral history one type of unicellular organism co-opted another, a photosynthesizing cyano-bacteria, and within the new host this became a chloroplast possessing the chemical molecule chlorophyll, which conducts the photosynthetic reaction. Since chlorophyll is a green pigment, plant life is over-whelmingly this hue.

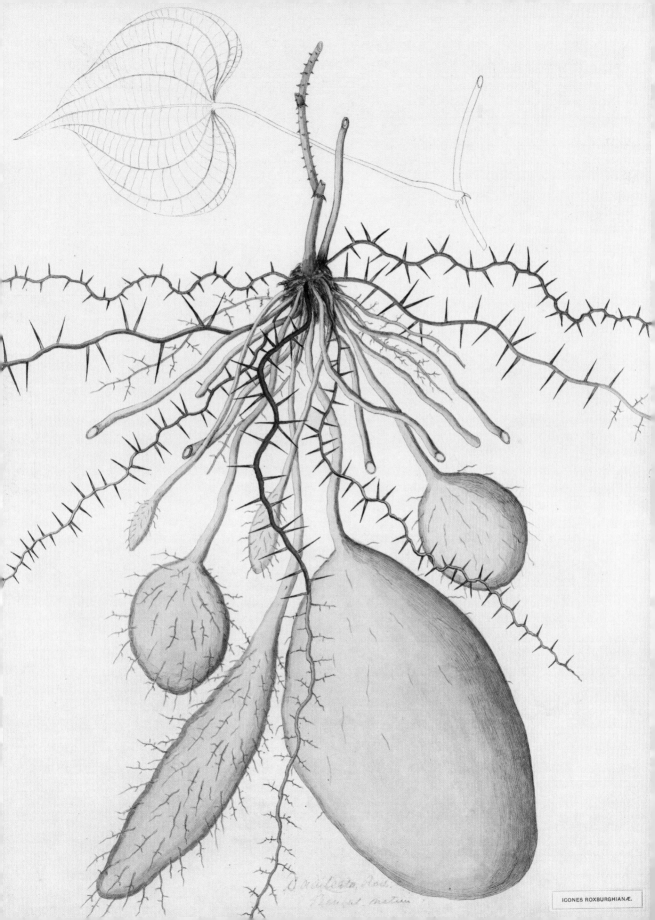

Bradleia Roa.
Bengal mature.

Above left After picking, tea leaves were once processed, dried and packed into bricks for transport from the tea estates. The tea bricks were often wrapped in bamboo for the journey, and were used as a form of currency and as gifts.

Above right Creating a new landscape: a plantation of Australian white gum trees (*Eucalyptus alba*) in the Bogor botanical garden, Java, Indonesia, 1894.

Through periods of time measured in immense geological epochs, all the different evolving forms of plant life proved to be much more than passive passengers on the revolving planet. They are now considered to be one of the active forces that have customized the atmosphere and surface of the Earth by their shaping and reprocessing of its fresh water and minerals, helping to make it more habitable for us. Some 290 million years ago some plants began to produce seeds, bringing the benefits of sexual reproduction. By 140 million years ago the first flowering plants (angiosperms) were evolving. In a relatively short period of geological time after this (some 60–70 million years) their forms and habitats had diversified to the extent that flowering plants were now dominant.

With flowers came colour. Green was now embellished with a dazzling array of tints. This was not for our delectation. Long before our love affair with them began, flowering plants co-evolved with other organisms to assist with the transfer of pollen, usually from one plant to another, for fertilization. While plants seem to have escaped the kinds of mass extinctions that saw the fall of the dinosaurs and the rise of the mammals, there have been important shifts in the world's plant populations. As the continental plates crunched more or less into their current positions about 65 million years ago and temperatures subsequently rose and fell to produce the successive ice ages of more recent geological

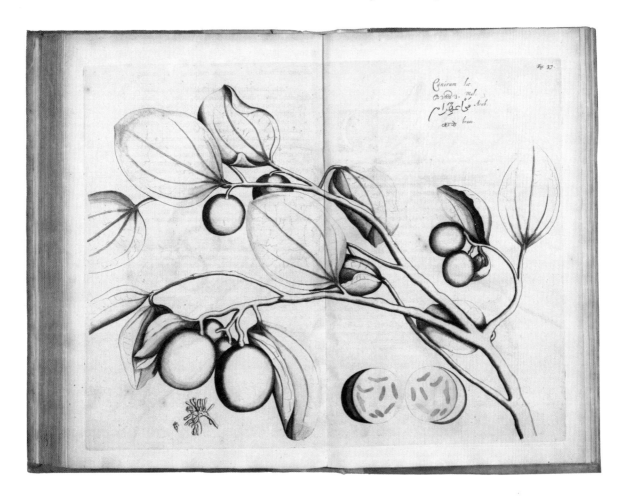

time, the grasses appeared to be winners and the forests losers, while arid land plants increased.

This was our inheritance some 12,000 years ago, as the last ice age ended, and this book is concerned with the diverse plant worlds of the planet as we made the transition from hunting and gathering to agriculture. How have we exploited some of these plants and what relationships have been forged between humanity and aspects of the plant world since then? In what ways do they impinge on our lives and we on theirs? Although each plant is allocated here to one of a series of categories, it is one of the wonders of the plant world that they can be highly efficient multi-taskers and could appear in several of our sections.

Our first, *Transformers*, looks at those plants that brought humankind to the settled way of life in many different parts of the world, including staples such as wheat, maize and rice. In *Taste* we explore plants that then enlivened and enriched our diet, from basic ones such as the useful alliums to the luxurious flavours of spices and saffron. *Heal and Harm* reminds us that there is often a fine balance

A nut-bearing branch of the Asian strychnos tree (*Strychnos nux-vomica*), from Hendrik van Rheede's *Hortus Malabaricus* ('Garden of Malabar', 1678–93). The plant has a long history of use in Ayurvedic medicine.

TULIPE

Top 'Tulipe' from *Les fleurs animées* by J. J. Grandville (1847). The gown is faithful to the passion for flowers infected with breaking viruses, which produce the stripes.

Above Sunflower (*Helianthus annuus*) by Pierre-Joseph Buc'hoz (1731–1807). This doctor-turned-botanist's artwork was inspired by Chinese and Japanese techniques.

Opposite Late 18th- or early 19th-century watercolour of *Bambusa balcooa*. Its sturdy stems can reach 30 m (100 ft) in height, making it useful for scaffolding and ladders in its native northeast India, Nepal and Bangladesh.

between those active substances in plants that can be either life-saving drugs or dangerous poisons in different doses. The featured plants and their products illustrate the breadth of different healing systems and the important plant base of much of the modern pharmacopeia. *Technology and Power* describes plants that have helped to create our material world – ships, houses, clothes and furniture, even weapons – all solid artifacts, both practical and pleasing.

The products of certain plants, such as tea, coffee, palm oil or rubber, as examined in *Cash Crops*, came to enjoy a global demand. Land was cleared for plantations of these crops and each altered a series of environments as surely as their products changed our patterns of growing, buying, trading, selling and consumption, and they are still potent shapers of world markets and fortunes. *Landscape* reflects on plants that seem to clothe parts of the world in unusual ways, even becoming emblematic – the massive redwoods of California, the eucalyptuses of Australia, the saline-tolerant mangroves of tropical coastlines. Each has a historical and contemporary role. For some the contemporary role also means contributing positively or negatively as invasive aliens.

Revered and Adored and *Wonders of Nature* allow us to celebrate the sublime and the amazing in the plant world, where mere utility slips into the background. Such plants have also shaped our history and our visual record of it. Date palms and lotus buds appear in the stone reliefs of Southwest and South Asia. Medical herbals by necessity required identifying pictures, which increasingly reached heights of artistic brilliance. The kind of visual beauty so evident in orchids, tulips and roses inspired artists in diverse cultures to capture nature's fleeting moments in perpetuity. So this book is lavishly illustrated with images of plants that celebrate their history, our history and the magnificence of their simply being there.

Bambusa
No. 1102 (Bambos) Baleooa Roxb.

Transformers
Settling Down, Tending the Fields

The shift from foraging to farming as the Earth warmed after the last ice age is one of the great transitions of human history. It was a protracted and by no means inevitable trajectory. The early farmers were often less well nourished than their foraging predecessors and subject to a disease burden resulting from close contact with humans and animals – intentional or otherwise. It is, though, from the permanence enshrined in farming that most of our present-day cultures have evolved.

Only in the continent of Australia did farming not take hold among the original inhabitants. In Asia, Europe, the Americas and Oceania it occurred independently, and involved different groups of plants – the 'founder crops' – grains, legumes, tubers; exactly 'what' depended on 'where'. Lentils, potatoes, yams, breadfruit, along with many others, formed the basis of diets around the world. Technology and seeds were then also carried from one place to another. Additional plants – fruits, vegetables, herbs – were brought into the garden. To the potent images made by humans as hunter-gatherers were added the art and artifacts representing a settled way of life. Storage, processing and cooking provided a new impetus to practical material culture. Deities believed to watch over the annual planting of seed and the subsequent harvest were feted and feared; life moved to new rhythms.

Farmers gradually altered the landscape, clearing and preparing land for growing crops and herding animals. Crucially, they also transformed the plants

VITIS VINIFERA varietas
Aleatico di Firenze

they came to rely upon. Under cultivation these plants were 'domesticated' so that they came to vary in fundamental ways from their wild ancestors.

Take the cereals. Archaeology has revealed that we gathered the seeds of many of these wild grasses such as wheat, rice and maize before we deliberately planted them. From purposeful cultivation of wild plants, early farmers went on to select those that had randomly mutated and better suited their needs. It was much more efficient to cut the stems of wheat or rice with intact ears and thresh them to release the grains than pick up what had been shed on the ground. A different genetic shift that resulted in seeds all germinating on the same environmental cue – spring rain or warm sunshine – was another great advantage over the wild state. Cereal growers also chose to plant seed of more flavoursome and plumper-grained parents that stood on thicker stalks, the better to hold up newly enlarged heads.

Seemingly ordinary, the founder crops of grains and pulses, roots and tubers were the transformers of our race around the globe. To their starch and protein are added here the fruit and oil of the olive and the flesh and juice of the grape, the elements, along with wheat, of the Mediterranean triad. They remind us that early farming encompassed a breadth of diet and technique. All represented a commitment to the land and an enduring sense of purpose. Farming was here to stay.

Above left A relative of the breadfruit of Oceania, the jackfruit (*Artocarpus heterophyllus*) is thought to have originated in India's Western Ghats and is used as a vegetable and fruit, even substituting for rice. It is an important food source in the subcontinent, especially for the poor, and is also attracting attention for increased utilization in the tropics.

Above right 'Aleatico di Firenze' grapes (*Vitis vinifera*). Aleatico grapes were known as a specific variety from at least the 14th century. Vines growing on Elba are used to make rich Muscat-like dessert wines, enjoyed by Napoleon Bonaparte during his exile on the island.

Wheat, Barley, Lentil, Pea

Triticum spp., *Hordeum vulgare*, *Lens culinaris*, *Pisum sativum*

Founder Foods of the Fertile Crescent

Ceres, most bounteous lady, thy rich leas
Of wheat, rye, barley, fetches, oats and pease.

William Shakespeare, *The Tempest*, Act 4, Scene 1

Wheat and barley, lentils and peas; these are the quintessential grains and pulses of the Neolithic revolution in Southwest Asia. They are some of the most important foods that allowed us to adapt our way of life in the face of the changing climate and plant landscape around 11,500 years ago at the beginning of the Holocene. Rather than being the brilliant inventors of a settled agricultural economy, we need to see our ancestors as coping as best they could by relying upon a small number of key crops when circumstances dictated and survival demanded.

The wild progenitors of these most important foundation crops (others included the still important chickpea and flax, and now forgotten bitter vetch) helped nourish those dwelling in the region long before formal domestication. Nomadic and semi-nomadic people were collecting wild wheat and barley, and grinding and baking the seeds from at least 23,000 BP ('before present'). After collecting, a long phase of pre-domestication cultivation followed (*c.* 14,500 to 10,600 BP), accompanied by hunting increasingly small prey. Fully domesticated crop lineages emerge from around 10,600–8800 BP onwards.

There was no single glorious moment when thrifty farmers somewhere picked out the best seeds from the previous harvest and effected a rapid change in either lifestyle or plant. Instead, there were a host of independent domestications, much reintroduction of wild species, and an exchange of domesticated varieties and know-how between people before farming could be said to have arrived for good. And if a different way of life suited, people would abandon farming and take up mobile pastoralism, using animals to process the plant cellulose that we cannot into milk and meat, or move on to new areas.

It was the storage potential and the surpluses generated from planned cultivation that fuelled the emerging agro-urban civilizations of Mesopotamia, Egypt and the Levant – the Fertile Crescent. Civilization in this case also required the abandonment of the egalitarian values of the limited size hunter-gatherer bands and the acceptance of a clear social hierarchy of haves and have-nots living in large cities, which measured their inhabitants in the tens of thousands. Large groups required administration and bureaucracy, and enabling technologies such as a

Chromolith. G.Severeyns. Bruxelles.

ERNST BENARY, ERFURT.

A. B. *Triticum Spelta* C. *Triticum polonicum.*
Spelt.   *Polnischer Weizen*

Blé de Miracle

Above left Both spelt (left) and Polish wheat (right) have large mineral- and vitamin-rich kernels and have enjoyed a resurgence as specialist grains. Spelt requires extra processing to remove the husk and Polish wheat tends to shatter, but they offer genetic potential to improve yields and disease resistance in the more commonly grown wheats.

Above right Blé de Miracle or cone wheat (*Triticum turgidum* subsp. *turgidum*) is an ancient free-threshing variety, domesticated around 10,000 years ago, probably in Southwest Asia, although the origins are obscure.

written language developed. At the same time, lengthy periods spent kneeling at the saddle quern and grinding for larger numbers than the family left their mark on the bones of knees and toes. We still marvel at the artistic and technological achievements, and monumental architecture facilitated by surplus grain. And we pore over the fragmentary remains of recipe books – clay tablets from 1700 BC – or tomb paintings for clues about eating like a Mesopotamian or an Egyptian.

It's hard not to think of farming to produce grain in abundance as inevitable, so integral to life are wheat (*Triticum* spp.) and especially the breads and pastas made from it. The domestication of two early kinds of wheat, einkorn (*T. monococcum*) and emmer (*T. dicoccum*), presented ancient cultivators with the opportunity to grow larger-eared plants with fatter kernels that mostly stayed on the stem, to be obligingly threshed on demand after harvest. Size matters, but the mutation resulting in a tough rachis (connecting the grain to the ear) that rendered the ears shatterproof after ripening was probably more important. It was even better when the threshing released naked grains such as those of the hard wheat (*T. durum*), preferred for pasta, and breadwheat (*T. aestivum*), rather than hulled grains that needed to be pounded in a mortar to remove the husk. Wheats are promiscuous natural hybridizers. The resulting hugely complex genomes from which we benefited are happy accidents of the three-million-year-old *Triticum* genus, although some are quite recent. Breadwheat arose about 8,000 years ago after two hybridizations, the second between a wild goat grass (*Aegilops*) and cultivated emmer wheat.

Barley (*Hordeum vulgare*) was used extensively for brewing as well as bread. Domesticated around the same time as einkorn and emmer wheat, the plump six-rowed barley appears about 8000 BP in early settlements. In the ancient economies

Lentils are cool season crops, and the bushy annual plants can cope with some aridity. The kind known as Puy lentils – 'vegetable caviar' – from the Haute-Loire region of France enjoy a Protected Designation of Origin status, recognizing both the variety and its locale, quite a feat for an ancient and humble legume.

An ear of barley, beautifully displaying its long awns – the stiff bristles extending from the husk of the grain. These photosynthesize and supply the developing seed with carbohydrates.

of Southwest Asia, barley was more important than wheat. It is hardier, coping better with cold and wet in part thanks to a post-domestication mutation making it resistant to powdery mildew. During successive cold snaps that disrupted life in the region, barley's resilience came into its own. What had once been a small-scale crop of natural rain-fed hill slopes, and river- and lake-sides, had become intensively farmed in valley-bottom fields that formed arable landscapes. Wheat retained its edge – it made tastier, finer bread. In better times barley was relegated to the food and drink of the poor.

As well as being used for bread and beer, grains were combined with the pulses in pottages. The wild ancestors of lentils (*Lens culinaris*) and peas (*Pisum sativum*) were probably weedy intruders in the cereal fields. But they more than made up for the land they occupied: with their amino acids, especially lysine in lentils, they added to the dominant grain carbohydrate. The retention of seeds and the disappearance of the requirement of a period of dormancy probably occurred between 11,000 and 9000 BP in the lentils' homelands of southeastern Turkey and northern Syria. Another improvement came with plants capable of growing upright rather than trailing along the ground; larger seeds appeared later. Domesticated peas (*c.* 8000 BC) also held on to their seeds, which became larger and lost their thick, rough seed coats, removing the obligatory dormant period before germination and making them more palatable.

The domesticated crops of the Fertile Crescent radiated out east and west as seeds and ideas followed trade routes and imperial pretensions. Peas were grown not to be eaten fresh, but to be dried for non-seasonal consumption. Able to adapt to harsher environments than lentils, they spread throughout Europe. Pea soup was an integral part of the Classical diet. The Greeks and Romans had their own versions, enriched with sausage and spice depending on taste and budget. Rome's imperial might, however, owed most to wheat bread. The vast imperial granaries of the port of Ostia were filled with grain from Sicily, North Africa and Egypt, requiring a dedicated fleet and navy to bring the harvest home. Lentils and peas found a new, appreciative home in the Indian subcontinent, the protein becoming especially valuable when vegetarianism took hold.

Wheat had helped usher in the Neolithic revolution and transformed the grass and woodlands of much of the world's flat lands into arable fields. In the 1950s and 1960s, along with rice, wheat was central to the Green Revolution. Again it was a chance mutation – a new response to the hormone gibberellin – that resulted in semi-dwarfing varieties with larger grain heads. Not only did these plants not topple over in wet or windy weather, but they also responded well to fertilizers with dramatic – up to six-fold – increases in grain yields. Transformers indeed.

Rice, Millets, Soybean, Grams

Oryza sativa, *Setaria italica* and *Panicum miliaceum*,
Glycine max, *Vigna* spp.

Asian Assets

Gin khao reu yung? [Have you eaten rice yet?]
Thai greeting

A severe flood in China forced the people into the mountains. When they returned home their plants had been washed away and food was scarce. As they struggled for survival, a dog came past with clusters of long yellow seeds in its tail. The people planted the seeds and after harvesting the crop their hunger vanished. This Chinese myth is one of many that reverences rice across monsoon Asia.

Rice feeds half the world today, and may well have fed more people in history than any other crop. Most of it is the domesticated Asian rice, *Oryza sativa*. West African *O. glaberrima* feeds far fewer, but intrigues plant breeders looking for new hybrids. In the story of Asia's staple crops, foxtail and broomcorn millets were to colder, drier northern China what the much thirstier rice was to the south. Each provided grains that could be foraged, cultivated as wild plants and then domesticated. As food and grain for brewing, rice would come to dominate and ultimately unite the palates of Asia.

The essential starch and vegetable protein combination played out in various configurations in Asia's emerging agriculture. In China, Southeast and East Asia, the soybean and its rich array of fermented products became a mainstay along with rice. In the Indus Valley flood plain, the Fertile Crescent package helped support the rich Harappan civilization. Further south, these crops of Southwest Asia made far fewer inroads. In India's south Deccan Neolithic it was the small green gram or Mung bean (*Vigna radiata*), black gram or urad (*V. mungo*) and horsegram (*Macrotyloma uniflorum*) and small-seeded millets (*Brachiaria ramosa*, *Setaria verticillata*) that formed a food complex. Subsequently, rice from the Ganges plains came to dominate. Mung beans are now the most universal of these foods: ground into flour for the dhosas or filled pancakes of southern India and, moving east, germinated as the bean sprouts common to much Asian cuisine.

Within the domesticated rice lineage the two most important groups are the *japonica* and *indica* types. Short, sticky *japonica* grains clump after cooking, making a more suitable rice to eat with chopsticks from a bowl than the longer, drier *indica* grains. As warmth and rain returned with the retreat of the glaciers the natural range of the ancestral rice plants expanded from their southerly glacial

DOLICHOS SOJA L.
Die Soja.

Oriza, Paddy, or Rice— Oryza Sativa, Hexandria Digynia

Calcutta 1817

Rice panicles or seed heads painted or collected by Mrs Janet Hutton in 1817 in Calcutta (Kolkata), where she lived with her husband, Thomas Hutton, an East India Company merchant.

refuges to cover a vast area of tropical and subtropical southern China, South and Southeast Asia. The perennial *O. rufipogon*, the wild ancestor of domesticated rice here, was collected and cultivated before the process of domestication was accomplished by 6,000 years ago. The resulting annual rice plants were heavier yielding (particularly when irrigated), with larger grains that were retained on the tall stalks ready for harvest. As with crops in the Fertile Crescent, the domestication process happened more than once across a wide area. Different traits may have been selected for in disparate places before being brought together and hybridized, including crossing with various wild types.

In the middle and lower Yangtze basins the classic paddy fields of wet rice production were developed (*c.* 4200–3800 BC) to mimic nature and provide an optimum artificial growing environment. These floodable fields were labour intensive, but their success in increasing yields drove rapid population growth. Paddies were bunded and then puddled, creating a hard pan to restrict water loss. Rice seeds were sown in nursery beds and transplanted by hand into the wet field. Help with some of these backbreaking tasks came from harnessing the water buffalo. In many parts of the wet rice-growing world little has changed. The later introduction of early ripening rice from the Mekong delta saw the extension of wet rice terraces on to slopes and repeat cropping in the lowlands.

From the initial centres of domestication the combination of rice plant and paddy field slowly reached most of the important areas of cultivation in Asia by

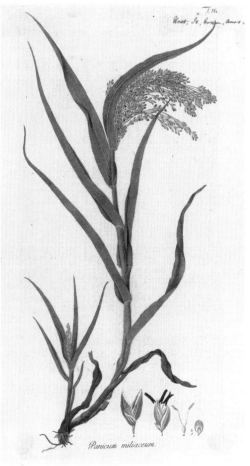

the Iron Age. There is an ongoing debate about the identity of the rice in South Asia. It is possible that another ancestral rice, an annual *Oryza nivara* type, was cultivated but not necessarily domesticated in the Ganges plain. These *indica* type rice plants were then hybridized with the incoming domesticated *japonica* types. The prized basmati rice, long assumed to be an *indica* variety, owes its delicate fragrance to a *japonica* ancestor.

Millets have remained important chiefly where other grains struggle. Ground to a flour, which is rich in protein and especially the B vitamins, millet nourishes, but even under optimum conditions cannot match the yields of other cereals. Initial collection and early cultivation of millets appears to have been around the Huang Ho (Upper Yellow River), but during a particularly dry spell (10,800–10,000 BP) millet farmers may have moved south to the Dadiwan region. Here and further east along the Yellow River, domesticated broomcorn (*Panicum miliaceum*) and foxtail (*Setaria italica*) millets with their larger seeds and better yields led to a series of settled, sophisticated cultures, although they did not

Millets: foxtail (*Setaria italica*, formerly *Panicum italicum*), on the left, and broomcorn (*Panicum miliaceum*) on the right. Millets are small-seeded cereals that belong to different genera. They are not closely related but do share similar growing needs. Heritage brooms in the US are usually made of another grass, *Sorghum vulgare*, which is thus also known as 'broomcorn'.

develop into city states. As it became more widespread, millet agriculture was most effective in tandem with the slowly expanding rice culture of the south. Along with wheat, after it had been introduced, the two millets, rice and soybean together became known as *wugu* – the five grains of China.

Soybeans (*Glycine max*) are derived from an ancestral vine common in northeast Asia, *Glycine soya*. Its early domesticators in Japan and along the Yellow River in China were faced with encouraging a recumbent vine to become an erect plant with larger seeds retained in the pod. Once domesticated, soybeans retained their ability to yield even when grown on poor soil. From the Zhou dynasty (1046–256 BC) the beans were cultivated and stored as an insurance against crop failure, but were little prized in their unprocessed state. This was in marked contrast to the growing use of fermented products – whole beans, sauces, pastes, relishes – and beancurd or tofu. Fermenting soybeans (the yellow variety in particular) increased palatability and nutritional value by adding vitamins and destroying the dangerous toxins that inhibited both protein digestion and iron and zinc uptake.

Soy products, already advanced in Han period China (206 BC–AD 220), became part of the wider regional cuisines of Korea, Japan and Indonesia. Though ubiquitous, and invaluable where meat was unaffordable or abjured, such products also had distinct local flavours, often allied to the specific microbes used in their creation. In countries without cheese and cured meat traditions, fermented soy could substitute. Soybeans are now grown on a huge scale in the Americas; much is for animal feed, raising concerns about land use and GM crops. And the rest? Like maize, soy is consumed unconsciously in highly processed foods, silently transforming our diet.

Opposite Mung beans and urad – or green and black gram – are now separate species. *Food-grains of India* (1886), from which this illustration of a Mung bean is taken, included both kinds of gram under the older name, *Phaseolus mungo*. It was, the author said, 'universally cultivated', 'highly esteemed' and 'resorted to by all, wherever possible, in times of sickness'.

Maize, Beans, Squash
Zea mays, *Phaseolus* spp., *Cucurbita pepo*
The 'Three Sisters' of the Americas

Modern maize was arguably man's first, and perhaps his greatest, feat of genetic engineering.

Nina V. Fedoroff, 2003

Above In 1712–14 Amédée Frézier travelled along the coasts of Chile and Peru to observe the 'genius and constitution of the inhabitants'. Here, in the foreground, he records the traditional method of grinding maize.

Opposite 'Runner beans' (*Phaseolus multiflorus* or *P. coccineus*) are one of the species of Central American beans brought to Europe after Columbus reached the New World. Initially they were grown as an ornamental for their striking red flowers, as is still the case in the gardens of North America. Varieties with white and bicolour flowers add to the mix.

Maize, beans and squash make up the 'Three Sisters' of American agriculture. The triad of crops was planted together in a method also known as the triple system, or 'milpa', from the Nahuatl word for 'at the field'. Milpa planting has many advantages: the maize, as well as yielding a product that is the basis of a nutritious diet, provides ready-made supports for the climbing beans, which in turn add dietary proteins (in the form of amino acids, the building blocks of proteins) that maize lacks, their roots also adding nitrogen to the soil; the squashes growing at their feet form a dense sea of green that keeps moisture and soil fertility in and suppresses weeds, while the fruits are an excellent source of carbohydrates.

The English word 'maize' comes from *mahiz*, meaning 'life-giving' to the Taíno-Arawakan people encountered by Columbus in the Caribbean islands on his first voyage. The alternative 'corn' comes from the German *Korn*. Maize (*Zea mays*) not only fed most of the ancient cultures in the Americas, it was also an important feature in many religious ceremonies and is the only plant to feature centrally in a creation myth: the Maya of Mesoamerica believed that the gods had created mankind from maize dough. The Spanish conquerors of the region were sometimes disturbed at the ease with which indigenous peoples could substitute maize bread and beer for the elements of the Eucharist. The Aztecs had their own deities associated with maize and the Maya a Maize God.

Genetic analysis has shown that maize's wild ancestor is teosinte, a grass that still grows naturally in western and southern Mexico. Teosinte plants produce few grains and these are actually indigestible to humans, but the stalk is sweet and was probably gathered to suck and to use in brewing. Domestication began in this area by 6,250 years ago, as occasional mutations led to larger and more easily gathered grain heads that were picked to be eaten and some of the seeds retained to be planted for the next crop, passing on the favourable characteristics. Maize had the essential advantage that it dries well and could be stored for later use.

The kernels could be used in a sort of gruel or ground into dough to make tamales and, later, the familiar tortillas. Such dishes were staples for centuries, and the stones on which the maize was ground and the flat griddles on which

Phaseolus multiflorus.

F. Guimpel 1ᵈⁱ.

Tab. 694

ALBUM BENARY
Tab XV

CUCURBITA PEPO L.
Die Gfebe

ERNST BENARY ERFURT

Above left Fruit, flowers, leaves and coiled tendrils of *Cucurbita pepo*. The plant has male (lower), pollen-producing, and female (top), seed-producing, flowers. The flowers, with their slightly sweet nectar taste, are highly regarded and can be stuffed and fried.

Above right Domestication of squash has produced varieties of this versatile plant displaying a great range of different colours and shapes, with equally diverse names, including bishop's mitre, elector's cap, turk's turban, la galeuse, angora and crookneck.

the tortillas were cooked are frequent finds on archaeological sites. Processing the harvest was woman's work: the kernels had to be removed from the cob and then ground. At some point, perhaps around 1500 BC, someone discovered the process of nixtamalization – soaking maize kernels in water to which lime or ashes have been added. The alkaline lime made the maize easier to process, but also had the added benefit of making the flour's vitamin, niacin, more available and so helped prevent pellagra (caused by vitamin B deficiency).

Domesticated maize spread from its Mexican homeland south to the Incas and north and east to the Native Americans. Inca rulers monopolized maize production and distribution, rewarding state workers who provided tribute in the form of labour with maize and the equally important maize beer, *chicha*.

All growers of maize discovered that planted alone, it quickly impoverished the soil. One solution, practised by the Maya, was regularly to clear new fields, often through slash and burn, leaving the old ones fallow for several years. But a widespread innovation helped soil fertility last longer: planting beans alongside the maize. As with many other legumes, beans fix nitrogen into the soil through bacteria living in nodules in their roots. The common bean (*Phaseolus vulgaris*), the second member of the 'Three Sisters', comes in numerous varieties, in many shapes, sizes and colours, including haricot, black, kidney and pinto beans. Beans were also grown throughout the American continent and, unlike maize, were

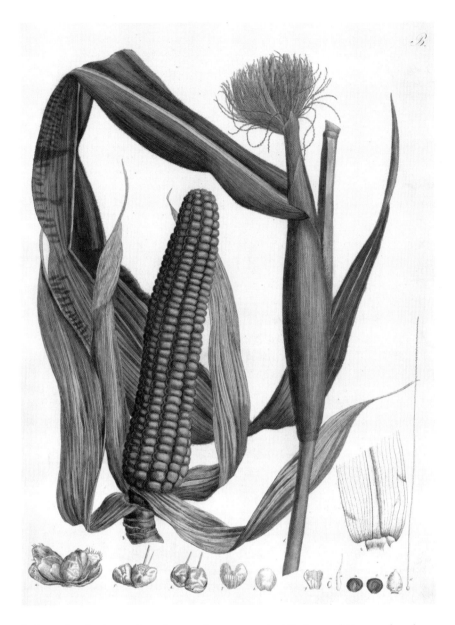

independently domesticated more than once – in Mexico and Peru and perhaps elsewhere. Wild beans almost 11,000 years old have been found in cave sites, but domesticated varieties are much younger. The beans were added to stews, providing valuable protein to the diet. Like maize, they could also be dried and store well.

The third 'sister' was squash, and its uses were recognized very early by American groups. In fact the bottle gourd (*Lagenaria siceraria*), inedible but prized for its durable hard-shelled exteriors, possibly found its way to the New World with early migrants crossing from Asia to America via Beringia. Dried shells of squash and pumpkins (*Cucurbita pepo*) could also be used as containers, but unlike bottle gourds these members of the squash family, Cucurbitaceae,

A ripe ear of maize, or *Zea mays*. An equal player in the historic 'milpa' triad, corn has become the basis of the industrial food chain thanks to its dominance in animal feeds and pervasive presence – in various guises – in our processed foods and drinks.

originated in America, not Asia. There is evidence of squash domestication in Mexican cave sites as early as 10,800 BP, which makes their domestication even older than that of maize. These earliest squashes had bitter, unpalatable flesh, but their seeds could be eaten either raw or roasted. Later varieties had flesh that could be added to bean and maize stews, or roasted on its own. Chillis, another domesticated American plant, were often included as well.

The 'Three Sisters' were all adaptable to a variety of climates and soils, and the system spread from Mesoamerica throughout the continent. Groups in the Mississippi River Valley started to grow maize on a large scale from about AD 900. The early English settlers learnt the system from Native Americans in Massachusetts, where it had been established by about AD 1000, and came quickly to value this New World grain and its companions. In fact the milpa system is still used, although planting has to be done by hand for the three plants to grow together well.

Maize, beans and squash each then spread around the world. Columbus took maize seeds back from his first voyage to the New World, and Spain, southern France and Italy soon had many corn fields, as the new grain's tastiness and versatility were quickly recognized. Beans, too, established themselves as mainstays of European diets (and provided the Moravian monk Gregor Mendel with one of the plants for his famous experiments on inheritance in the 19th century). Although squash had the problem of storage, the seeds at least could be kept over winter. Squashes have proliferated into a great range of colours, shapes and sizes, from diminutive to the giant pumpkins (*C. maxima*) grown competitively.

While all three are still major elements of people's diets globally, it is perhaps maize that is most significant and is now one of the most important crops grown worldwide. Significant work to improve yields by hybridization was done in the early 20th century, especially by Donald F. Jones in America. As well as for straightforward human consumption, maize is now grown for oil, corn syrup, animal fodder and biofuel. Its genetic makeup has been thoroughly investigated (the plant has more genes than human beings have) and it was one of the first to be successfully genetically modified, to improve resistance to disease and pests as well as to augment yields. Barbara McClintock's work on gene regulation and 'jumping genes' was performed on the maize chromosome, which she mapped, helping us understand the historical changes the maize plant has undergone, winning her the 1983 Nobel Prize for Physiology or Medicine.

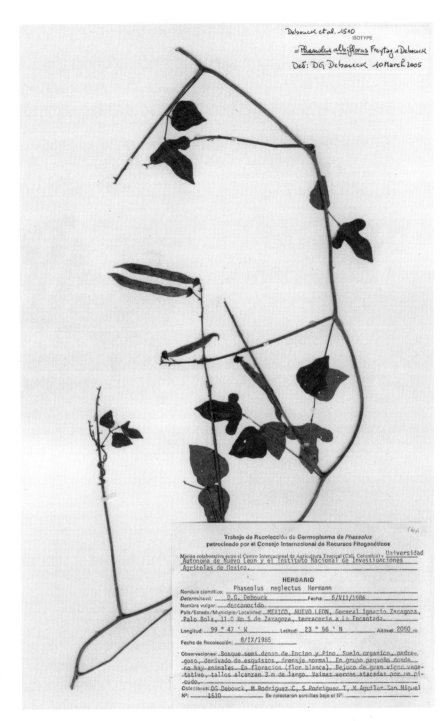

Debouck et al. 1510
ISOTYPE
of *Phaseolus albiflorus* Freytag & Debouck
Det: DG Debouck 10 March 2005

Trabajo de Recolección de Germoplasma de *Phaseolus*
patrocinado por el Consejo Internacional de Recursos Fitogenéticos

Misión colaborativa entre el Centro Interacional de Agricultura Tropical (Cali, Colombia) y Universidad Autonoma de Nuevo Leon y el Instituto Nacional de Investigaciones Agricolas de Mexico.

HERBARIO

Nombre científico: Phaseolus neglectus Hermann
Determinavit: D.G. Debouck   Fecha: 6/VII/1986
Nombre vulgar: desconocido
País/Estado/Municipio/Localidad: MEXICO, NUEVO LEON, General Ignacio Zaragoza, Palo Bola, 11.0 Km S de Zaragoza, terraceria a la Encantada.
Longitud: 99 ° 47 ' W   Latitud: 23 ° 56 ' N   Altitud: 2050 m
Fecha de Recolección: 8/IX/1985
Observaciones: Bosque semi denso de Encino y Pino. Suelo organico, pedre-goso, derivado de esquistos, drenaje normal. En grupo pequeño donde no hay animales. En floracion (flor blanca). Bejuco de gran vigor vege-tativo, tallos alcanzan 3 m de largo. Vainas verdes atacadas por un pi-cudo.
Colectores: DG Debouck, M Rodriguez C, S Rodriguez T, M Aguilar San Miguel
Nº: 1510   Se colectaron semillas bajo el Nº:

A herbarium (dried and pressed) specimen of *Phaseolus albiflorus*, held at Royal Botanic Gardens, Kew. Wild beans such as this one, collected at altitude in Mexico, may provide valuable genetic material for improving domesticated beans in the face of climate change.

Potato, Sweet Potato, Groundnut, Quinoa

Solanum tuberosum, Ipomoea batatas, Arachis hypogaea,
Chenopodium quinoa

South American Heirlooms

Potatoes, by feeding rapidly growing populations, permitted a handful of European
nations to assert dominion over most of the world between 1750 and 1950.

W. H. McNeill, 1999

The Inca culture rose quickly, flourished for a brief time and collapsed catastrophi-cally, the victim of disease and Spanish aggression in the 1530s. These Andean peoples created a vast and impressive empire, and while they inherited the domes-ticated plants they consumed, it was through them that the world learned of several foodstuffs that now enjoy global use.

One of them, the potato (*Solanum tuberosum*), has proved so resilient that it is now grown in almost every climate. There is evidence of exploitation in the Andes from at least 5000 BC and it was certainly well established in the diet in the high-lands by 2500 BC. Andean potatoes have diversified into a wide variety of shapes, sizes and colours, and Peruvian markets still reflect this. Europeans first encoun-tered the potato in 1537, but it has had a rocky road to world dominance. The tubers were tried in Spain and Italy, but not much liked, and in any case didn't thrive in southern European climates. Northern Europeans initially regarded the potato with suspicion, some Protestants rejecting it because it is not mentioned in the Bible. Also, the plant belongs to the same family as the deadly nightshade, so was thought to be poisonous.

The early South American varieties in Europe, accustomed to their home-land near the Equator, matured when days and nights were roughly the same length. In northern climes this meant autumn, when the risk of frost is high. Potatoes hybridize easily, however, so during the 18th and early 19th centuries, European horticulturalists worked to develop new varieties that were more suit-able to European and North American climates. They were encouraged in this work because the plant is a heavy cropper and provides lots of calories. To over-come the daylight problems, stocks from Chile, further south from the Equator, were developed, and they found favour in northern Europe, first in France and Germany, and then in Ireland, where farmers with only limited land could now feed their families.

The notorious potato famine of 1845–49 actually began in Belgium, but its consequences were most catastrophic in Ireland, where starvation killed around a million, with the same number emigrating. 'Potato blight', which devastated the

bridges sale 1873

Arachis   *China gram*

Irish harvests, and other diseases are still troublesome, but new potato varieties are more resistant. Late 19th- and early 20th-century breeding extended further the climatic areas in which the plant could be grown – this is now almost worldwide. China, India and the US are very large producers, but whether baked, boiled, mashed or chipped, potatoes feature widely in cuisines everywhere. The tuber itself is nutritious, although its preparation may make the end product less so.

Sweet potatoes (*Ipomoea batatas*) come from a completely different family, the common name 'potato' deriving from early European confusion. 'Batatas' was the West Indian name for the plant Columbus's men first encountered in Haiti in 1492. The sweet potato is the tuberous root of a tropical vine, related to the morning glory flower, and probably originated in the region from southern

After the yellow flowers of the groundnut are fertilized a specialized structure known as a peg grows downwards into the soil. The tip absorbs water and nutrients, and stimulated by the lack of light will develop into the pods containing the nuts. The vines and leaves are used as high protein 'hay', while the empty shells can be used to make particleboard.

La Pomme de Terre

Lat: *Solanum Tuberosum* Allem. *Grundbir*, Angl. *Potatoe*, Amerie *Papas*.

G. de Nangis del et Sc.

Mexico to Venezuela. Wild species were consumed possibly as early as 8000 BC and evidence of them is found in early archaeological sites in Peru. Although modern sweet potatoes contain high amounts of starch and some sugars, the original tubers were more fibrous than sweet. Nevertheless, they spread very widely in pre-Columbian times, north through Mexico, the southern parts of North America and the Caribbean islands, and, more surprisingly, throughout the Pacific islands and into Australia and New Zealand.

How this latter spread occurred, probably as early as 2,000 years ago, has generated much speculation. Since the tubers could not survive in seawater long enough to float to these places, it was probably through human agency, whether deliberate or accidental, although birds carrying seeds is a possibility. Whatever the mechanism, sweet potatoes were being grown in Oceania and the Antipodes long before Europeans arrived there. Also perhaps surprisingly, *Ipomoea* initially found more favour in Europe than its *Solanum* cousin, and was grown and consumed in the Mediterranean countries before Spanish and Portuguese ships spread it to Africa and Asia. In West Africa and many Oceanic islands, the sweet potato began to compete with the yam (*Dioscorea* spp.) as a dietary staple. Sweet potatoes are used in many of the same ways as 'Irish' or 'white' potatoes, and in sweeter dishes. China, where the sweet potato was introduced in the late 16th century via the Philippines, is now the world's largest producer and consumer, although it is also grown in many parts of Africa, in India and the southern United States.

Another Peruvian plant that enjoys wide cultivation is the groundnut (*Arachis hypogaea*). In English this is often known familiarly as the 'peanut', which is at

Above A version of the traditional Andean plough with a hardened wooden or metal tip – the clods of earth it threw up were then broken up with a hammer. Such simple tools were used successfully to cultivate potatoes for thousands of years high up in the Andes.

Opposite A flowering and fruiting potato plant. The French pharmacist Antoine Parmentier, a potato promoter, encouraged Queen Marie Antoinette to wear potato flowers to add social cachet to the tuber at the end of the 18th century. After fertilization seeds are produced in the small ball-shaped fruit. If planted, these may give rise to new varieties that can then be bulked up and propagated through their tubers.

QUAMOCLIT PATATE.

The sweet potato, painted by Jean-Théodore Descourtilz from drawings made by his father, Michel Étienne, during his sojourn in Haiti and travels in the Caribbean. Much of M. Descourtilz's collections were lost in the Haitian revolution, but some survived and were used to prepare the illustrations for the *Flore pittoresque et médicale des Antilles* (1821–29).

least partially accurate since, while not a nut, the plant is a legume (like the pea). Other popular names include earthnut, goober and Virginia peanut, the last after the most common American variety. Groundnuts, which develop underground, have been found in archaeological sites dating to between 6500 and 4500 BC, and were eaten raw, roasted or ground into a paste in the New World before the Europeans came. They were already grown by the Taíno in Hispaniola by 1492, and were taken from there to the Philippines and East Indies by the Spanish. They then spread to China and Japan. The Portuguese took them from Brazil to Africa, where they quickly became an important crop, and also to India. Although they have long been grown as fodder for livestock, groundnuts are popular with humans, raw or roasted, and also pressed into cooking oil. They are central to Malaysian cuisine and are processed into a paste in the United States (and elsewhere) known as 'peanut butter'.

Unlike the three cosmopolitan foodstuffs described above, the principal indigenous grain of the Andes, quinoa (*Chenopodium quinoa*), has penetrated the rest of the world less successfully. The plant's large, spinach-like leaves are edible, but it is mostly the tiny white or pink seeds, the 'grains', that are eaten. The seeds must be processed in an alkaline solution before use to remove toxic saponin, a compound that protects the plant from birds. Quinoa was a major Inca food crop, since the plant can survive frosts and grow in poor soil and at high altitudes. It was used in stews, ground into a flour for bread or tortillas, featured in religious ceremonies and could form the basis of Inca beer, *chicha*. It continues to be extensively used in Peru, Bolivia and other Andean countries, but is still mostly an 'exotic' elsewhere. Cultivation and consumption may well increase, given the fact that the seeds are rich in the protein lysine, vitamins and minerals and are regarded as a 'super crop'.

'White' quinoa featured in *Curtis's Botanical Magazine* in 1839. The editor apologized to his readers for the plant's lacklustre appearance, but reminded them that along with 'handsome plants' the magazine featured those of 'peculiar interest'. Quinoa deserved its place as the 'chief nourishment' of the temperate regions of South America. The Incas revered it as the 'mother' of grains.

Sorghum cernuum.

Sorghum, Yams, Cowpea
Sorghum bicolor, Dioscorea spp., *Vigna unguiculata*
Staples South of the Sahara

There is nothing that captures the 'soul' of soul food more than black-eyed peas.
Lindsey Williams, 2006

A grain, a tuber and a legume: each originated in Africa, spread far afield but remain mainstays of many African diets. A grass, sorghum (*Sorghum bicolor*) may have been cultivated in Ethiopia, Sudan or Chad as early as 4000–3000 BC. Almost certainly domesticated from a wild subspecies, *S. verticilliflorum*, it can reach as tall as 4 m (13 ft), with a head full of seeds and has the advantage of growing on dry soils. Its value was quickly recognized; it spread to India by about 2000 BC, and was adopted in many parts of Africa by the early Christian era. Its early cultivation is central to debates about the origins of agriculture in Africa, a process that is still not entirely understood. Sorghum remains the most important grain grown in sub-Saharan Africa, and is fifth in world grain production.

Fermentation helps break down sorghum's proteins into more digestible and usable molecules, and the grains are then very adaptable, as flour, pastes or added to stews or porridge; they can also be used to make beer. There are several races, or types, of *S. bicolor*. Although these freely hybridize, each has different characteristics and growth requirements, helping to establish them in various parts of Africa as a principal grain. (Sorghum almost always competes with other grains, especially millet but also maize, depending on soil quality and annual rainfall.) One type, durra, the most drought tolerant, has tended to spread with Islam. The second, kafir, is common south of the African Equator; its high tannin content helps protect it from birds, but also means that processing is necessary before human consumption. The third type, guinea, prefers higher rainfall and is the favoured grain in West Africa. There are also several wild strains that can be harvested in times of famine. In the early spread of sorghum cultivation in Africa, the grass also helped support livestock tending.

India produces sorghum in considerable quantities, where it is used both for human and livestock consumption. It is also well established in the United States – a variety with a high saccharine content has been exploited to produce a cheaper substitute for maple syrup, although its major use there is as fodder.

Yams (*Dioscorea* spp.), several closely related species of the same genus, are also an important African food crop. They are tropical vines, although they can

Opposite Like many African edibles, sorghum was used to feed slaves during the 'Middle Passage' from West Africa across the Atlantic to the Americas. It was thought that providing familiar foods would increase their chances of surviving this horrific journey.

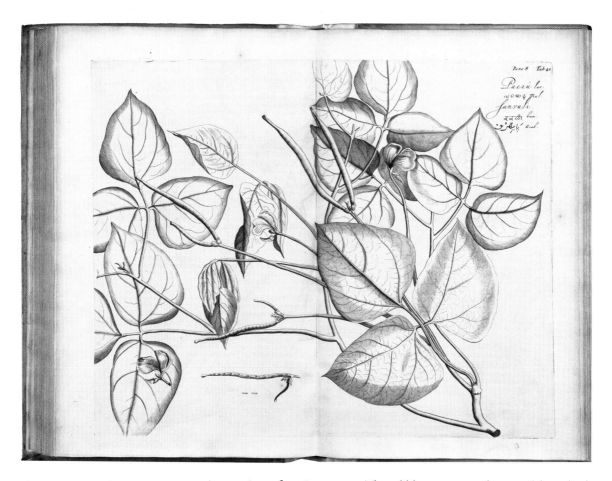

The cowpea spread from its probable site of domestication in West Africa, arriving in southern India around the 2nd millennium BC. It expanded with the advent of better tools and irrigation methods. Local varieties were selected for and it became an important rotation crop after rice. Hendrik van Rheede would have found it widely grown as he surveyed the Malabar Coast in the 17th century and featured it in his *Hortus Malabaricus* (1678–93).

grow in a variety of environments. The edible parts are tubers, and have both advantages and disadvantages. Among the former is the fact that, once dug up, the tubers can be stored without refrigeration for several months, thus affording food in times when other alternatives are scarce. They can also be very large (one has been weighed at 60 kg/132 lb), and, properly tended ('ennobled'), are extremely productive. Their main disadvantage is that the tubers are often found deep in the ground (up to 2 m/6 ft), making harvesting backbreaking and difficult. (Some Asian species produce their tubers above ground.) This aspect has featured in debates about African agriculture for periods when iron tools were unknown and stone hoes would have been the instrument of choice. Yams probably were transported for planting elsewhere by migrating groups as early as 2000 BC and would have been dug up by hunter-gatherers long before. Most yams must be cooked before eating to remove harmful toxins.

Although yams evolved in many parts of the world, the two principal African species are *D. cayenensis*, dominant in West Africa and with yellow flesh, and *D. rotundata*, closely related (and generally today considered as merely a variety of the former), with white flesh. They provide a staple food throughout central Africa, although they now compete with the sweet potato and other imported plants. Their flesh can be grated and pounded into a paste, or cooked in a variety of ways. Yams are central to the diets in many Oceanic islands, and in India and China, where *D. alata* is the most common species. Asian yams were introduced into Africa and Madagascar in the 1st millennium AD, and then taken to the New World consequent to the slave trade.

Another African staple also crossed the Atlantic in European ships: in the 17th century the Spanish introduced the cowpea (*Vigna unguiculata*), known in the United States as 'black-eyed peas'. Also a plant that requires heat to mature, black-eyed peas are still part of southern American cooking, and a central feature of 'Soul Food'. Most are allowed to dry, but the fresh, less mature pods can also be eaten, generally cooked with pork and chilli to give them more flavour. Cowpeas are also still widely used in Africa and Asia, as well as throughout the Caribbean islands; in Haiti they are a principal food crop.

As a legume, cowpeas fix nitrogen, so are useful co-planters, especially in Africa, where they form part of a mix of basic agriculture. In India they can be substituted for lentils in making dahl, and are often part of bean curries. A particular variety, *V. sesquipedalis*, produces very long beans, and although most don't live up to their name, 'yard-long bean', a few do. These are generally eaten as a pod rather than a pea.

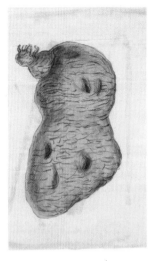

Dioscorea cayenensis drawn in Sierra Leone in 1921. In the Yam Belt of West Africa there are elaborate yam festivals and yams may have contributed to the formation of settled village life and agriculture. Tending garden plants to maturity and crossing with occasional wild forest plants (a continuous process) created today's 'ennobled' yams.

Each breadfruit is a compound fruit or syncarp, which develops from the female inflorescence. The male inflorescence is bottom left. Breadfruit have sustained the peoples of Oceania for thousands of years; there are over 2,000 vernacular names for the many cultivars found there.

Taro, Breadfruit

Colocasia esculenta, Artocarpus altilis

Fuelling Oceania

[Breadfruit] is baked entire in the hot embers, and the inside scooped out with a spoon. I compared it to Yorkshire pudding.

Alfred Russel Wallace, 1869

Taro (*Colocasia esculenta*) and breadfruit (*Artocarpus altilis*) could hardly be more different. The one is a root (actually a corm), with large, attractive leaves, the other a tree that can reach 85 m (278 ft). But they share an important similarity – their products are delightfully starchy. Their histories are also closely intertwined, as they have both long been staples in the Pacific islands. 'Taro' is in fact also the name used for three other edible corms of different genera, but with many of the characteristics of *Colocasia*.

Colocasia taro originated in South Asia, probably India or Burma (Myanmar), but spread some 6,000 years ago to Thailand, Malaysia, Indonesia and the Philippines. It was soon established in Papua New Guinea and then on islands throughout the Pacific. In Hawaii there were 70 local names for it. The plant produces few seeds, so its spread by propagation has been due to human intervention. This is achieved by transplanting the top of a root plus a little portion of stem, which made it possible to replant and eat from the same root. In this way, farmers could always have them on the go, in different stages of growth. From planting to harvest takes from 9 to 18 months. A tropical plant, taro requires plenty of rain or irrigation, and though it can be grown in drier conditions, it thrives in flooded areas, which means that it began to compete with rice cultivation in many places.

Taro cultivation centred on the tropical Pacific islands, but it was also carried west, to Africa, Egypt, some Mediterranean islands, and, by the 8th century, into the Iberian Peninsula. It eventually made its way to the Caribbean and South America. Because the root has such a high starch content it was welcome everywhere at a time when sweet things were in short supply. It is easily digested and acquired a reputation as a baby food and good for adults with digestive problems. It retains an affectionate place in Oceanic cuisines, although the availability of other, cheaper foodstuffs means it is often reserved for special occasions.

Breadfruit (*Artocarpus altilis*), related to the mulberry, was probably indigenous to what is now Papua New Guinea, although it had been taken to many Pacific islands long before Europeans intruded on the scene. It can yield each year as many as 150–200 fruits as large as a good-sized melon, which, when cooked,

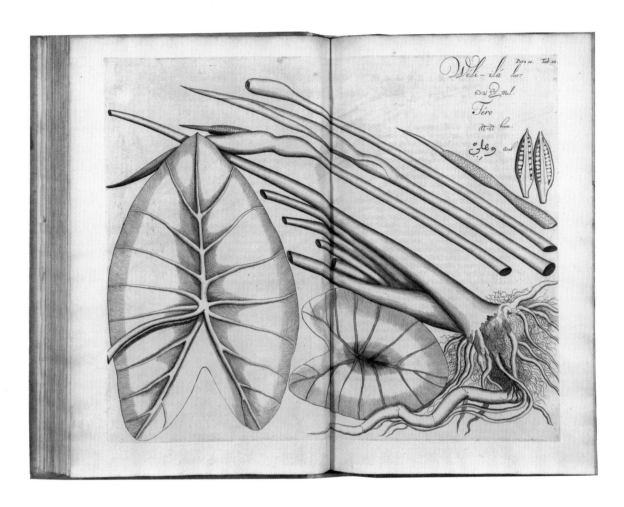

produce a pleasant, sweet substance. In addition to being baked in pits lined with hot stones, it can also be dried and the flesh ground into flour. The fermented fruit is sometimes made into cakes, and it can be interchanged with taro in making a sweet porridge. As a bonus, the tree's timber can be used in construction. There are now as many as 200 varieties, with modern, seedless ones predominating.

The tree's productivity and easy availability made it popular throughout the region, and it much impressed Europeans when they saw it. The naturalist Joseph Banks, sailing with James Cook in the 1770s, determined to spread it through tropical North America. Breadfruit was the object of the journey of HMS *Bounty*, scene of the legendary mutiny in 1789. William Bligh, the ship's commanding lieutenant, survived and later returned to gather breadfruits for transplanting in the New World, where they were intended to provide cheap food for slaves. Although the experiment hardly provided the anticipated panacea, the tree has been established throughout the Caribbean and in tropical South America.

Taro, from Hendrik van Rheede's *Hortus Malabaricus*. In addition to the edible leaves and central corm (from which the leaf stalks arise), he illustrated the flower structures and fruit with seeds. Natural flowering and seed production are rare events, but this reluctance can be overcome, unlocking the potential for crossing different varieties of this important food crop.

Alfalfa, Oat

Medicago sativa, Avena sativa

Speed the Chariot and the Plough

Lucerne is by nature an exotic to Greece ... it having been first introduced into that country from Media, at the time of the Persian wars with King Darius; still it deserves to be mentioned among the very first of these productions. So superior are its qualities.

Pliny, 1st century AD

Though whimsical, this ploughing scene from Pietro de' Crescenzi's *De omnibus agriculturae* (1548) serves to underline the importance of the continued development of agricultural equipment and fodder crops.

Domesticated animals are a great boon for many purposes, but they, too, must eat. The earliest named fodder crop is alfalfa or lucerne (*Medicago sativa*). This small-seeded clover-type legume was part of the same steppe biome of western Eurasia that also brought the domesticated horse. Horses of the nomadic tribes could graze on naturally growing alfalfa, and it became the preferred cultivated fodder crop of those who adopted the horse for trade and warfare.

The Anatolian Hittites, acclaimed early charioteers, recorded on clay tablets overwintering their horses on alfalfa. The idea spread along trade routes and with invading armies. Darius I of Persia fed his camels and livestock with it, and exposed the Greeks to this fodder crop. Alfalfa became part of the Roman war machine. Columella and Pliny in the 1st century AD applauded it, providing detailed instructions on field preparation and cutting regimes. Interest came from the East too. The 2nd-century BC Han emperor Wu, keen to consolidate his territorial gains, sought the famed horses of the Ferghana valley (eastern Uzbekistan), and their food. The route from the Han capital Xi'an through Ferghana led on to the Black Sea and became the northernmost of the Silk Roads.

Alfalfa was not only a useful crop which could be cut and stored, as a legume it also improved soil quality and could feed the subsequent crop planted in a rotation system. Its benefits seem to have been forgotten in Europe before returning through Spain, perhaps with the Muslims. The Italians referred to its Spanish and Hungarian origins during the Renaissance. But by this time another fodder crop had become important to farming, especially in northern Europe.

Wild oats (*Avena sterilis*) originated in the Fertile Crescent, but unlike wheat and barley remained essentially a weedy intruder in domesticated fields. In this way oats spread throughout Europe. In the northwest, when the weather was poor, oats tended to outperform other cereals. The next step was single crop cultivation, with evidence from Germany about 4,000 years ago. Domesticated oats (*Avena sativa*) followed a thousand years later. Oats became a staple cereal in Celtic Europe, reinforcing the Romans' disdain for a crop they thought fit only for barbarians and animals.

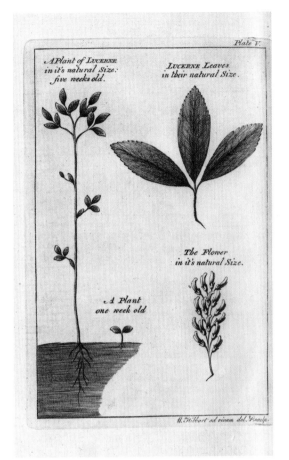

The ploughs of Southwest Asia and southern Europe only had to cope with light soils. North and west of the Danube and Alps soil was heavier, and often recently cleared of trees. The sturdier wheeled plough with its mouldboard, capable of cutting through and turning the sod, was initially drawn by oxen. Faster horse-power was limited by the existing yokes. Improvements to yoke design and the introduction of the horseshoe released the full potential of the medieval plough. And high-energy oats fed the horse. Oats were combined as part of a rotation system, but their ability to perform better in more marginal soils than wheat and barley was a bonus.

Beyond Europe, oats continued as animal rather than human food, although their cholesterol-lowering soluble fibre has increased appreciation. Alfalfa travelled to the New World in the 16th century with the Spanish and Portuguese. From South America it went north. Seed introductions from the original Asian homelands and selective breeding yielded new varieties. In the 21st century alfalfas are cleaning up pollutants including long-lasting herbicides via the micro-organisms that live in association with their roots and break down the chemicals. A very modern use for an ancient forage.

Above left Walter Harte studied farming methods on the continent while accompanying a young pupil on his Grand Tour of Europe. In the second of his *Essays on Husbandry* (1764) he addressed the 'Culture of Lucerne by Transplantation'. He was a keen advocate of this 'handsomest of all the grasses which are (improperly called) artificial'.

Above right Two varieties of oat as depicted in J. Metzger's *Europaeische Cerealien* (1824). These are now regarded as synonyms rather than distinct plants.

Olive

Olea europaea subsp. *europaea*

The Quintessential Oil

The olive tree is the first of all the trees.

Columella, 1st century AD

Olive trees are synonymous with the Mediterranean. Tree, fruit and oil have helped to define the land, climate and people living around the Middle Sea. While domestication wrought some changes, olives still are most productive where the Mediterranean's pattern of temperature and rainfall, with warm, wet winters and hot, dry summers, is replicated. Whole fruit contributed to the kitchen and table, but it is the oil that has always offered so much. As lamp fuel it burned bright and clean, lighting up the darkness. Its unguent properties soothed cracked skin and calmed sunburn, glossed the hair and made excellent soap. Olive oil is a natural solvent; it melds the flavours in a dish and suffused the wearer with the glorious perfumes of the ancient world. In the eastern Mediterranean olive oil became a sacred anointing liquid, and wherever the region's peoples and ideas travelled, reverence for the oil went too.

Once simply a vital source of calories, olive oil now embodies the 'good life'. It is the quintessential ingredient in the much-vaunted 'Mediterranean Diet' believed to prevent cancers and atherosclerosis. Thanks to the complex of mono-unsaturated fatty acids and polyphenols, there is in fact much more than nostalgia in a generous drizzle of extra virgin olive oil; the very name of the oil's highest grade conjures a sense of exceptional purity.

The wild olive (*Olea europaea* subsp. *sylvestris*) of the Mediterranean is more thorny shrub than tree, with wider leaves and smaller fruits than the cultivars it has produced. The muted greens, greys and blacks of wild olives blended with other evergreens to create the typical maquis vegetation of the limestone slopes of the Mediterranean basin. Today the wild plant is vastly outnumbered by domesti-cated cultivars and feral olives. Hillside olive groves may originally have involved clearing away competing vegetation and managing existing trees, as well as plant-ing new ones. Pollen analysis from the eastern Mediterranean shows just how dominant olives were during the zenith of cultivation in the region from *c.* 550 BC to AD 640, but the olive's importance to humans can be traced much further back.

Archaeological evidence indicates that the semi-nomadic hunter-gatherers who lived on the shores of the Sea of Galilee were collecting large quantities of

Opposite A flowering olive branch – an international symbol of peace and friendship. Since olives are drupes or fruits, olive oil (like palm oil) is a 'fruit juice'.

wild olives from around 19,000 BP. By the Late Neolithic the fruits were not merely collected but also crushed for oil. Sites of the following Chalcolithic period, inland and upland from the coast, show production was expanding. Whether these were wild or cultivated olives, or a mixture of both, is debated, but exploitation was intensifying. The transition from wild bush to farmed tree, currently thought to have taken place in the northern Levant, would have been a slow process. The continuing selection for larger, fleshier, oilier fruits involved repeated interchange between wild and cultivated trees, and had to be done in a species that is slow to come to maturity and fruit.

Olive trees are very long-lived and can be productive for several hundred years if well managed, greatly adding value to the land. These traits reinforced the strong historical association with the settled way of life; a grove was worth defending. During the later Bronze Age, olive oil, increasingly produced from cultivated trees, became a valuable and much traded commodity. High-quality oil demanded, as it still does, skills in crushing, extraction and transport, the last aided by the burgeoning pottery industries. Where olives were less easily grown, such as in Mesopotamia or Egypt, the oil commanded high prices among elite customers.

What had begun in the Levant would spread more widely with the Greeks and Romans. Rome's taste, and thus its need for olive oil was colossal. Romans also began to eat the fruits at table (suitably prepared by soaking and brining to combat their natural bitterness). Romans planted and often irrigated groves in North Africa, southern Italy and Andalucía. With typical efficiency they also developed a grading system for oil and a bureaucracy to guard against fraud and adulteration (a continuing problem). Such was its popularity that the emperor Septimius Severus (r. 193–211) added oil to the *annona* food subsidy that helped keep Rome's citizens quiescent.

Today, olive groves are celebrated for their sustainability. They are rich in insect life and provide important feeding places for resident and migratory birds. Olives not only cope with, but actually do better on, poor soil. Waste from olive pressing yields a useful fertilizer, animal feed and a further source of fuel in addition to the trees' wood. Their expansive, shallow roots help bind the soil on slopes. Where these slopes have been stepped into the iconic terraces, each level area helps hold rainwater and slows degrading run off. Since irrigation also increases a tree's yield, farmers benefit too. Olives can be unreliable croppers with a tendency towards a biennial pattern of rich and poor years, but proper pruning helps counteract this. Much harvesting is still done by hand; green olives are picked earlier than the riper black ones.

In return for its bounty, the olive was transformed into a symbolic object. It was the tree gifted to Athens by Athena. Olive oil from the sacred trees, the *moriai,*

Olive

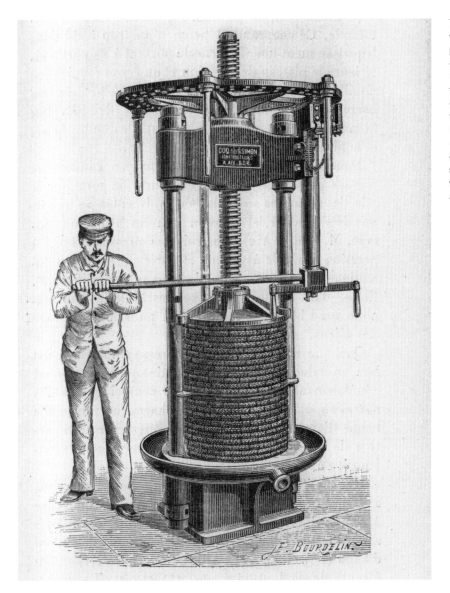

To extract the oil, the olives were crushed and the resulting paste pressed, with the released oil and water then separated. Pressing was made easier after the introduction of hydraulic equipment in the 19th century; this French model is featured in P. d'Aygalliers's *L'olivier et l'huile d'olive* (1900). Today integrated centrifuges prepare the oil in sealed systems, limiting contact with the air and treating the fruit as gently as possible.

was given as prizes at the Panathenaic games, as the olive wreath went to the victor at the Olympian games. The olive branch has transcended its Greek origins, and biblical and Talmudic connotations, to become an international symbol of peace and reconciliation.

ci ueteres seruati querendi sunt, quia meliores durabilic huc semimaturi existunt.

De tempore uindemie,   Cap.   :

uuæ ma
ideo sub
manens
tardius t
neas sed
us & m
gnoscii
gustu &
trius & Africanus sex solū dies habere uuā, & non plus
enim granū uuæ iam nō uiride sit, sed in ea nigredine ut
dem esse debet secundū naturā generis illius uuæ, signi

Grape
Vitis spp.

In Vino Veritas

The vine [bears] three kinds of grapes, the first of pleasure, the next of intoxication, and the third of disgust.

Anacharsis, 6th century BC

Above Wine was traditionally matured and transported in wooden, usually oak, barrels – vine and tree combining to produce the desired flavour.

Opposite Cornichons blancs, also known as Ladies' Fingers from their characteristic elongated shape, were prized table grapes. Belgian artist Pierre-Joseph Redouté, best known for his flowers, also applied his stipple engraving technique to produce these sumptuous fruits for *Choix des plus belles fleurs et des plus beaux fruits* (1827–33).

It may perhaps be a particularly English cliché to take grapes to a patient in hospital, but there is a certain logic. Red and purple grapes contain large quantities of flavonoids, plant pigments that provide the attractive colouring and function as antioxidants. The role of these molecules in preventing the devastating diseases of modernity – cancers, cardiovascular disease, neurodegenerative illnesses, rheumatoid arthritis – has not proved to be as simple as first suggested, and as dietary supplements antioxidants have disappointed, but there is still something in fruit eating that does have positive effects. Grapes are delicious, healthy little packages, with a handy trick on their skin.

The once widespread wild Eurasian grapevine (*Vitis vinifera* subsp. *sylvestris*), a temperate liana, grows in riverine woods. The trees provide support for the woody twining plants as they grow upwards to the light of the canopy. Separate male and female plants produce flowers, and pollination leads to fruit on the females. The grapevine was domesticated in the mountainous areas of the northern Zagros, eastern Taurus and Caucasus. It was part of the second wave of domestications, following the grains and legumes. The key was a hermaphrodite plant occurring in the wild, which required only a single genetic mutation, and its selection by humans.

Grapevines strike easily as twigs simply stuck into the ground; in this way desirable variations could be easily reproduced and undesirable traits avoided, since the seeds of grapes produce highly variable plants. It was then possible to plant a grove or terrace of clones and create a new landscape – the vineyard. Such simple but highly effective vegetative propagation was followed later by the more complex technique of grafting – a vital step in the history of plant control. Thereafter, careful tending, supporting and pruning yielded grapes for consumption fresh, for boiling down into sweetening syrups and for drying as raisins, sultanas and currants, which became storable staples of European winter cuisines. The leaves made wrappings for various dishes and the prunings made good firewood for cooking.

Long before domestication the small, sharp-tasting wild grapes were collected and their innate predisposition to ferment presumably enjoyed. As the fruit ripens

Cornichons blancs.

and bursts, the juice comes into contact with naturally occurring yeasts on the skin and alcohol is produced. The next step – named the 'Palaeolithic Hypothesis' for the period in which it may have occurred – argues for active grape gathering and fermentation into wine. It cannot be determined with any accuracy because the necessary artifacts have not survived. Evidence of the possible use of wild grapes (or hawthorn berries) in alcoholic grog comes from the Neolithic site of Jiahu, Henan Province, north-central China, and at Hajji Firuz in the northern Zagros a pottery jar from 7,400–7,000 years ago had been used to store grapes or grape products, but grape domestication and winemaking proper are thought to date from 5,500 to 5,000 years ago. From the mountainous regions to the north, production spread to Mesopotamia and then Egypt. Wine and raisins became important traded goods that were stored in amphorae lined with pine resin, the resulting taste today recalled by Greek retsina wine.

In all these cultures, as for the Greeks and Romans to come, wine was closely associated with ritual and religion as well as sociable drinking. Besides the common bar, the elite Greeks had their *symposion*, the Romans the *convivium*, and the cult of Dionysus/Bacchus flourished. Wine was also central to the Jewish Sabbath and the Christian Eucharist. In the unsettled conditions of Europe after the fall of Rome, wine production remained closely tied to the Church. As the monastic orders – especially the Benedictines and Cistercians – expanded, well-tended vine-yards followed in some of our now revered wine-producing areas, though secular wine makers played their part too.

As a medicament in the system based on the four humours, wine was endowed with hot and moist qualities. It was particularly valued in nourishing the emaci-ated and the cold and dry elderly – easily digested, wine fired up declining vital spirits. And wine, like man, matures to old age, when it oxidizes to vinegar. Wine vinegars were not just wastes, but carefully produced condiments and preserva-tives. They were also vital during the plague as they were thought to contract the body's passages such as the pores of the skin and nostrils, thus helping prevent entry of the noxious vapours or miasmas believed to cause the disease.

Grapevines accompanied Europe's voyages of exploration and colonization. From Mexico, *vinifera* vines spread to Peru, Chile and Argentina and northwards to California. In South Africa, the Cape of Good Hope was particularly recep-tive, though it took until the 1840s to find the right terrain in Australia and New Zealand. As so often happens, man's movement of plants around the world also spread disease. Three significant grape parasites – powdery mildew or oidium (*Erysiphe necator*) and downy mildew (*Plasmopara viticola*), and an aphid (*Viteus vitifoliae* previously *Phylloxera vitifoliae*) – originated in the northeastern USA. Grapevines here had co-evolved with the pathogens and were rarely as severely affected as the imported ones, which is why vines taken to the east coast from the

A View of Scalanova near Smyrna —

From 1700 to 1702 French botanist Joseph Pitton de Tournefort undertook a botanizing expedition to the Levant, travelling inland as far as Georgia. Claude Aubriet, a renowned botanical artist, accompanied him and illustrated the places, plants and people they encountered. Here at Scalanova near Smyrna (now Izmir) on Turkey's Aegean coast, they watched a new vineyard being planted in this raisin- and wine-producing region.

16th century onwards had failed, prompting the breeding of cultivars based on the local vines.

Copper sprays controlled oidium, but in a bid to produce resistant plants an exchange of cuttings in the 1860s probably introduced the root-eating aphid to Europe, where it ravaged the French wine industry. Further attempts by the growers to hybridize their way out of trouble introduced downy mildew to Europe. The phylloxera devastation was stopped by grafting *vinifera* vines on to resistant native American rootstocks. The technique that had helped establish better vineyards in antiquity had saved the global wine industry.

Hybridization has produced new grape varieties that have introduced different flavour notes and broadened the palate, particularly for younger wines. Older wines have become high-end commodities or collectors' items, seemingly never to be drunk. There are perhaps 10,000 grape cultivars today, including many seedless grapes ideal for drying or eating fresh, even if this means the fruits serve no purpose for their parent plants – the ultimate transformation.

Taste
Beyond the Bare Necessities

The staple crops were closely bound up with the founding of civilizations, but while they may feed and sustain us and satisfy our hunger, alone they do not always fulfil our cravings for flavour, something to make the ordinary special. It is this additive quality of plants and their products that is explored in *Taste*.

Pre-eminent is the gorgeous spice saffron, redolent of the expensive and the exclusive. This of course has much to do with availability. One man's ordinary is another's exotica. From India and the Spice Islands came pepper, nutmeg and cloves; the desire for these condiments drove trade and some of the great voyages of exploration. In contrast the alliums may seem a rather ordinary part of the kitchen garden. Garlic, onion, shallots and leeks each have their own cocktail of pungent sulphur compounds, leeks being the most delicately flavoured. Garlic is one of today's superfoods, but its smell made it unattractive and there was a distinct class bias. Its consumption was encouraged for Roman slaves and soldiers, who were thought to benefit from its strengthening powers.

The Romans also held asparagus in high esteem. They did much to bring this plant into cultivation and valued its medicinal properties. When diet was more about the qualities of ingredients and less about food groups such as starch and proteins, it was prized in the Renaissance for its subtlety and easy digestibility. Brassicas are often considered humble greens, but what astounding versatility

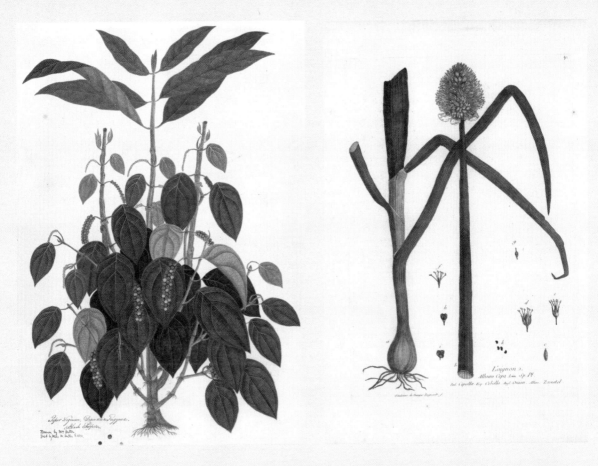

in these highly nutritious species and cultivars. The ancestors of today's cabbages (and from them the cauliflowers, sprouts, broccolis and kohlrabi) were headless, the heads appearing first in plants raised in northern Europe. At last, a key plant from the cold north. In China a plethora of oriental leaves belong to the same family.

Beer, of various kinds, often seen as an indulgence or a danger today, was an important part of the daily diet of labourers in early civilizations. It provided a source of clean water, calories and pleasurable intoxication, but it didn't keep. Enter the hop. However, it was only in the 8th or 9th century in Europe that this aphrodisiac herb was added as one of several flavourings, and its preservative qualities subsequently determined.

Two tastes originated in the New World, but became so essential in their adoptive homes that they now help define their cuisines. Chilli peppers were embraced as soon as the Iberians carried these Mesoamerican delicacies home and onwards to their colonies to the east. And tomatoes, with their sweet, sharp fruits, had no equivalent in the Mediterranean cooking that they have come to dominate. Initial suspicion gave way to great affection for the 'love apple' from the 17th century.

Above left Watercolour of a black pepper plant (*Piper nigrum*) painted or collected in India by Mrs Janet Hutton in the early 19th century.

Above right An onion plant, including flower, from *La Botanique* (1774) by the husband and wife team Nicolas-François Regnault and Geneviève de Nangis-Regnault.

Saffron
Crocus sativus
The Spice of Conspicuous Consumption

Strands of dried saffron from the pharmacy, though they are identical in all respects to those found in the spice jar. Cilician doctors recommended saffron to the Egyptian queen Cleopatra to ensure her complexion remained unblemished.

His floure doth first rise out of the ground nakedly.
John Gerard, 1636

Crocus sativus is a small, rather unprepossessing corm that gives rise to delicate purple flowers with distinctive large, dark orange, almost red, dangling stigmas. These stigmas – plant parts otherwise rarely used – produce one of the world's most expensive food commodities: the spice known as saffron. Since its incorporation into diet, dyestuff, mythology and medicine, golden-yellow saffron has been highly desired. From the saffron-dyed robes of the Buddhists to the saffron-strewn streets of ancient Rome greeting emperor Nero, its association with the sacred and the elite is long-standing. Yet, used whole or ground, its pungent taste, insistent smell and strong colour mean only a little is needed.

In its Persian homeland saffron enhanced traditional rice dishes such as pilaf and shola. The Phoenicians traded it along the seaways of the Mediterranean, taking it as far as Spain in the west, even if the Muslims reintroduced it during the heyday of their empire there. The Mughals expanded its use in India. Crusading knights and pious pilgrims reputedly brought corms back from the Holy Land, though there's no evidence for this method of introduction. However it arrived, saffron began to be grown in Italy, France and Germany. Walden in eastern England prefixed Saffron to its name as large-scale production took hold in this wool town.

The medieval cooks of Europe regarded saffron as an essential part of the courtly aesthetic, which ostentatiously used food as a display of wealth. Fashions changed in aristocratic households, but saffron remained a key ingredient of Provençal bouillabaisse and Milanese risotto. Swedish lussekatter bread made on the feast of Santa Lucia testified to its rarified integration into the store cupboard there. High cost and scarcity invited adulteration. Safflower (*Carthamus tinctorius*) flower parts could be used to mix in with whole stigmas, while the powdered root of turmeric (*Curcuma longa*) would substitute for the ground spice. The Germans moved to suppress this sleight of hand in the 15th century – taking it seriously enough to make it a capital offence, punishable by burning or burying alive.

The cost is due to the methods of harvesting and processing, which are still unmechanized. Each plant has three flowers, opening on successive days; each bloom has three of the precious stigmas, the long, string-like female parts that provide the glorious spice. After the whole flower has been cut, the stigmas – joined at the base – must be separated and dried. It takes 70,000 flowers to yield 1 lb (0.45 kg) of dried saffron, and these 70,000 flowers take up approximately a tenth of an acre (404 m^2). Picking of the flowers is best done before sunrise, when the moisture content is at its optimum. In the saffron fields of the world's largest producer, Iran, harvest occurs in October and November over a 20-day period of intense activity.

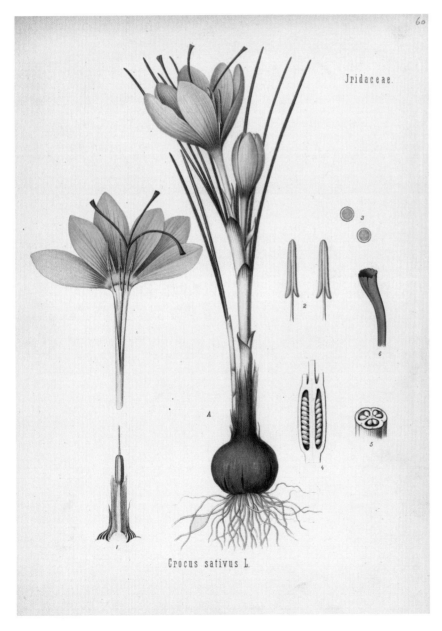

Crocus sativus L.

Both drug and condiment, saffron was prescribed to cut thick, unhealthy humours such as those thought to congest the lungs in cases of consumption. Similarly, it was included in recipes where its subtle quality would balance the effects of other ingredients and counteract the production of sluggish humours in the healthy body.

Tab. XXV

Planck. Jc.

PIPER NIGRUM. L.
Der schwarze Pfeffer.

Nutmeg, Cloves, Pepper

Myristica fragrans, Syzygium aromaticum, Piper nigrum

Riches of the Indies

The isles Of Ternate and Tidore whence merchants bring Ther spicie drugs.

John Milton, 1667

So important were spices to early modern Europe that the indigenous sources of two of them, nutmeg and cloves, were called the Spice Islands. There were even Spice Wars, as Portugal, Spain, the Netherlands and Britain vied for control of these valuable commodities once direct sea routes to the east were discovered. Traditional land routes had allowed spices to be laboriously transported to Europe even in antiquity, and early Mediterranean cultures enjoyed cinnamon, nutmeg, cloves and pepper. But they were expensive because they were costly to transport, generally first by boat using the monsoon tides of the Indian Ocean to the ports of the Arabian peninsula, and then by land and sea to final destinations. Nutmeg could be had in Rome in the 1st century AD, although no one was quite sure where the precious spice came from.

In fact, the nutmeg tree was indigenous to only a few islands in an archipelago east of present-day Indonesia, the Banda, or Spice Islands. This evergreen tree is unique among spice-bearing plants in that it produces two spices as well as a fruit. Nutmeg is the prized kernel of the fruit and the even more expensive mace is the thin covering around it. The fruit itself was eaten, often candied in honey or sugar syrup. The laden tree with its golden fruit was described in the 16th century as 'the loveliest sight in the world'. Nutmeg was long traded in the Orient, valued in China and India, and in Constantinople (present-day Istanbul) in the Byzantine period. As well as flavouring food it was also used to sweeten ale and to keep clothes fresh smelling. In China it was thought to have medicinal properties, and Europeans also believed it alleviated a number of ailments, including bubonic plague, heightening demand.

After the Portuguese navigator Vasco da Gama opened a direct ocean route to the Orient in the late 15th century, spices from the area became more readily available. Even though the distance of sailing around the tip of Africa was twice as long as the various land-and-sea spice routes of antiquity, it was cheaper to carry them by water (and pay duty only once). Ships could call at the Banda Islands to take on nutmeg, and then at nearby Ternate and Tidore in the Moluccas for cloves, the dried unopened flower buds of *Syzygium aromaticum*.

Above Fortunes were made from black pepper. On 16 November 1665 the diarist Samuel Pepys described the spices in the hold of an East Indiaman as the 'greatest wealth [lying] in confusion that a man can see in the world'. It seems almost possible to smell the 'Pepper scattered through every chink, you trod upon it; and in cloves and nutmegs, I walked above the knees; whole rooms full.'

Opposite A flowering and fruiting stem of the black pepper plant (*Piper nigrum*). The detail shows the brightly coloured ripe berry, the dried black peppercorn and, with the coat removed, the white peppercorn.

Right A flowering sprig, fruit, nut and seed coat or aril of the nutmeg. On its home soil on the Banda Islands in Indonesia the trees bear blossoms and golden fruit all year round. The fruit splits to reveal the vivid scarlet aril surrounding the nut, and this dries to the familiar yellow-brown, brittle mace of the kitchen.

Opposite The aromatic leaves and flowers of the clove tree. It is the unexpanded flower that is traded as the spice, but the purple berries that form after flowering can be preserved in sugar and eaten as a post-prandial digestive.

Having loaded their holds with the dried spices, which kept well, they could return (shipwreck or piracy permitting) with a cargo worth a fortune. Cloves were widely sought after in Europe, to keep the breath fresh as well as for flavouring food and for use in medicines. There is probably no spice that has not been used medicinally.

The profits to be made in trade in the two spices were so great that Spain, and then the Netherlands and England, competed with Portugal for the monopoly. Spain was briefly successful before the Dutch East India Company (VOC), formed in 1602, established a strong presence on the Spice Islands. The East India Company of London, which received its royal charter in 1600, was only moderately successful in its attempts to supplant the Dutch, although the small island of Run, one of the nutmeg group, became (briefly) England's first overseas possession. The English were not able to hold on to it, and eventually the two nations reached an agreement that gave the Dutch their Spice Islands monopoly, but obtained New Amsterdam (and therefore Manhattan and other Dutch possessions in North America) for the British.

The English East India Company subsequently concentrated on India, where the French and Portuguese were already involved, and which was the home of another major internationally traded spice – pepper (*Piper nigrum*). This vine grows naturally in south India and was cultivated for local consumption and

Black pepper from the *Hortus Malabaricus* ('Garden of Malabar', 1678–93). Conceived and overseen by the naturalist and colonial administrator Hendrik van Rheede of the Dutch East India Company, these innovative volumes recorded and illustrated over 700 plants of the Malabar region and were in effect the first flora of Asia. The Latin text was augmented with plant names in local languages – Kankani, Arabic and Malayalam.

export to other parts of Asia and to Europe. As a vine it needs support: wild plants often hitched up a coconut tree, but posts were employed in plantations. The berries were harvested in early autumn and then dried, which coincided with the monsoon winds that aided the homeward journey of European ships.

In Classical times the Greeks and Romans prized pepper, and knew both black and white versions. White pepper is merely the black form picked when riper and with the outer layer removed. Roman authors complained that their balance of payments was in danger, since gold was the only commodity they had to pay for the large quantities of pepper they consumed. The Greeks also used what was called 'long pepper' (*Piper longum*), a hotter version that was always more expensive and which they mistakenly assumed came from the same plant. Long pepper, indigenous to northwest India, was still esteemed in Europe until the 17th century, and even two centuries later, cooks using Mrs Beeton's classic cookbook (1861) would have needed it to hand. It lost its standing in Western cooking with the greater availability of the American chilli pepper, which provided the same heat and could be cultivated in Europe.

Black pepper was the most widely traded spice, and worldwide demand seemed insatiable. Even in the Middle Ages it was fairly readily available in Europe – if at a price, given the long transport involved. Menus often mention it, and its medicinal properties were as highly prized as its enhancement of food. It was thought to reduce phlegm, warm the body and manage flatulence.

Control of the production of all three spices became as important as their transport and trade. The Dutch were ruthless in the Spice Islands, at one point destroying the entire crop of nutmeg trees on Run to retain their monopoly, unfortunately thereby also reducing genetic variability. Botanical espionage is a long-standing practice around the world, and both nutmeg and cloves were smuggled out of their home islands for transplant elsewhere. Pepper plants had already been exported from India to Indonesia before the beginning of the Christian era, which meant that European ships could add pepper to their consignments of nutmeg and cloves after the sea route to the region had been established.

Although nutmeg, cloves and pepper all require a warm climate, globalization and demand have ensured that they are now grown commercially in many countries. The more widely a spice is grown, and the greater number of powers controlling its production, the more difficult it is to regulate its price on the world markets. Grenada in the West Indies has rivalled Indonesia in nutmeg production, and there are pepper plantations in Malaysia and Brazil as well as India. Cloves were introduced into Zanzibar in the 20th century, but a fungal disease largely wiped out the industry there, leaving Indonesia, the spice's original home, as the main world producer as well as consumer, where ground cloves are mixed with tobacco in cigarettes.

Above A natural necklace using nutmegs and cloves as 'beads', part of the rich holdings of the Kew Economic Botany Collection.

Left Foliage, flowers and fruit of the nutmeg (*Myristica fragrans*). Although indigenous to the Moluccas and Banda Islands in the southern Pacific, this group was painted in Jamaica by Marianne North. Among the islands of the West Indies, Grenada – 'Isle of Spice' – dominated production (second only to Indonesia) until Hurricane Ivan (2004) devastated the island's plantations.

Capsicum baccatum was probably domesticated in Bolivia, although its natural range stretches from Peru to Brazil. It is known as aji in South America, where the subtle bouquet and distinct flavours of the cultivated varieties are much valued.

Chilli Peppers
Capsicum spp.
Some Like It Hot

[The Aztecs] have one [plant], like a pepper, as a condiment which they call chilli, and they never eat anything without it.

'The Anonymous Conqueror', *Narrative of Some Things of New Spain*, 16th century

Few plants pack such a potent punch as the fruit of the medium-sized bushes *Capsicum annuum* or the even hotter *C. frutescens*, the chilli (also 'chili' or 'chile') used in Tabasco sauce. They owe their heat to an alkaloid called capsaicin, which probably evolved, as did many other plant alkaloids, as a protection against predators. It is most concentrated around the seeds; this is why removing the core and seeds of the chilli takes away much of the heat. Since capsaicin is not soluble in water, drinking it does not help relieve the hot sensation.

C. annuum was found in the wild in Mesoamerica and was prized by the Aztecs. The stews of the milpa fields (maize, beans, squash) would have been tastier with it, and it was added to chocolate. Chillis (the word is from the Aztec language Nahuatl) were picked from the wild as early as about 7000 BC and were cultivated by about 4000 BC. By the time of the Spanish conquest they came in various sizes, shapes, colours and degrees of hotness, and had also spread to North America and the Caribbean islands. Columbus encountered them on his first visit to the New World; their use by the Taíno people of Santo Domingo confirmed his mistaken belief that chillis were the pepper (*Piper nigrum*) of the East Indies and that he had found the western route to those islands. Although the two are unrelated, the word 'pepper' has stuck as one common name of *Capsicum*, especially for the larger, sweeter varieties.

Columbus' physician Diego Álvarez Chanca took chillis back with him to Spain, where they were immediately popular with a few. They proved easy to cultivate there, much to the alarm of merchants of the lucrative trade in black pepper. European sailors soon spread the new taste sensation to Asia, Africa and Brazil. Chillis quickly became so important in Indian cooking that it is hard to realize that they are a relatively recent addition. The Renaissance naturalist Leonhart Fuchs even assumed they were native to India.

As with most new additions to the diet, chillis were also evaluated as a medicine. They were regarded, like black pepper, as hot and dry, and so useful against cold, wet diseases. The chilli never really established itself as a mainstay of medical therapeutics, however, and it is as a condiment that the plant is cultivated today, with Mexico and India the major producers. More than a dozen major varieties are

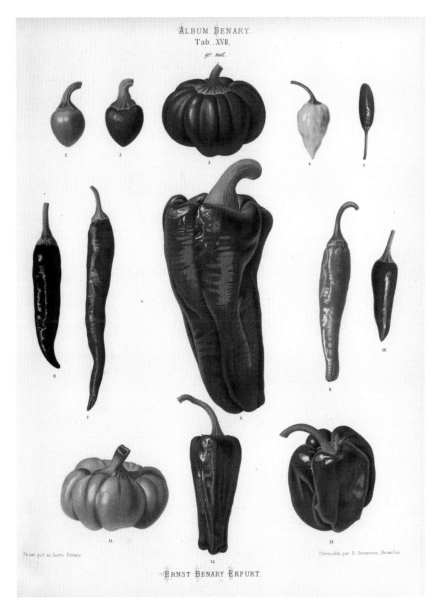

ALBUM BENARY.
Tab..XVII.
gr. nat.

Ad nat. pict. in horto Benary.

Chromolith par O. Severeyns. Bruxelles

ERNST BENARY ERFURT.

Something of the rich variety of shapes, sizes and colours of cultivated capsicums can be seen here. Their heat is measured in Scoville Heat Units (SHU) on a scale devised by the American pharmacist Wilbur Scoville in 1912. Bell peppers score 0 SHU as they have no capsaicin, while long red cayenne scores 30,000–50,000 SHU. The range reflects the interaction of varietal genetics and the circumstances under which the plant is grown.

available, with dramatic differences in flavour and hotness. Most modern ones are derived from *C. annuum*, but the seeds of *C. frutescens* are used to make cayenne pepper. Paprika (the word is also derived from *Piper*) is a mainstay of Hungarian cooking, and also comes in a variety of degrees of heat and shades of sweetness.

In addition to *C. annuum* and *C. frutescens*, other species of capsicum have local importance, including *C. chinense*, which is popular in the West Indies. Despite its name, *chinense* did not come from China, nor is the dominant species, *annuum*, an annual. When Carl Linnaeus named it in the 18th century, he was responding to its European behaviour; in tropical countries, capsicums are perennials.

Garlic, Onion, Shallots, Leek

Allium spp.

Hellfire and Brimstone?

And, most dear actors, eat no onions nor garlic, for we are to utter sweet breath.

William Shakespeare, *A Midsummer Night's Dream*, Act 4, Scene 2

Opposite A range of varieties of European onion (*Allium cepa*), characterized by skin colour, form and season of readiness.

Cut and consume the familiar kitchen alliums and the result is distinctive. Onions (*Allium cepa* var. *cepa*), shallots (*A. cepa* var. *aggregatum*; *A. oschaninii*) and leeks (*A. porrum*) may induce tears, and all, as well as garlic (*A. sativum*), can leave a lingering smell on the breath, in sweat and in urine. These well-known effects are due to the high concentration of organic sulphur compounds found in many of the 800-plus species of the genus *Allium*. Sulphur is the ancient stuff of hellfire and brimstone, a fitting evocation of the alliums' potent qualities.

Intact plant tissues bind the sulphur into stable compounds (cysteine sulphoxides), but these become volatile (primarily as thiosulphinates) when the cells are compromised, for instance by cutting or chewing, and the sulphur comes into contact with an enzyme called alliinase. With onions, which make us cry the most, the volatile sulphur dissolves into sulphuric acid in our eyes' natural moisture. These processes have evolved to protect the plants from predation. We might not like the tears, but the sharp, biting tang of the raw and the wonderfully sweet mellow flavour of cooked alliums have been appreciated for millennia as upfront vegetables and rounded-out background seasonings.

Alliums are northern hemisphere plants (only two species are native to the southern hemisphere). The diversity of these edible plants, with species stretching from the dry subtropics to just below the Arctic Circle, afforded plenty of opportunities for foragers. The greatest number is concentrated from the Mediterranean basin through Central Asia, Afghanistan and Pakistan, the same region that was probably home to the main domesticated garden plants. East Asia also provided some cultivars including bunching onions (*A. fistulosum*) and Chinese chives (*A. ramosum*).

Onions are thought to have been cultivated first in Central Asia, though the wild form is not now known. Leeks look likely to have come from the area of the eastern Mediterranean and western Asia. Kurrat leeks (*A. kurrat*) were cropped for their leaves by the 3rd millennium BC and the stemmed forms (*A. porrum*) followed. The garden origins of garlic in Central Asia date back to around 3000 BC. Nomadic tribes probably then took garlic into Mesopotamia and India. As with

M.... pinx. in horto Benary.

Chromolith. G.Severeyns. Bruxelles.

ERNST BENARY, ERFURT.

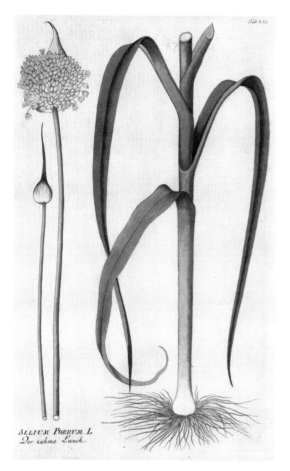

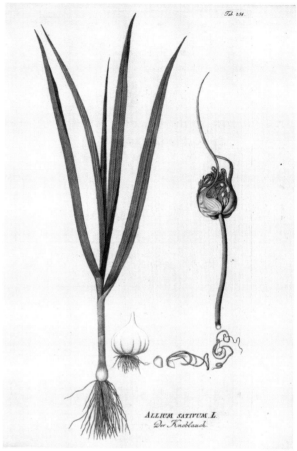

onions, the dry bulbs were easy to carry and even simpler to plant – individual cloves can be gently inserted straight into the ground.

The Egyptians likened the layers of the onion bulb to the concentric spheres of their cosmos. This may have been why they placed onions within the body cavities of mummies, offered them on altars and swore their oaths on them. Egyptian attitudes towards alliums presaged much of their later culinary history. Garlic, onions and leeks were part of the daily fare of Egyptian workers, and according to the Bible the Israelites while being led by Moses through the wilderness from bondage in Egypt lamented the alliums of their captivity. Consumption was not restricted to the lower classes, but it is said that garlic breath would lead to exclusion from the Egyptian temple by its priests. Alliums – especially garlic – were both delicious and polluting.

The Greeks and Romans also appreciated garlic for its strengthening qualities. It was the daily food for their labourers, athletes, sailors and soldiers, who needed vitality and stamina. For the same reason it was fed to fighting cocks before they entered the pit. Garlic was widely employed both externally and internally as a medicament against a range of illnesses. In part this was due to its demonstrable anti-microbial properties, although it would not have been understood in this way at the time. Garlic's easy availability, cheapness and broad application led the physician Galen (129–*c.* 210) to refer to it as '*theriaca rusticorum*', the poor man's cure-all, an appellation that lasted as long as Galenic humoral medicine.

Like the Egyptians, the Greeks and Romans worried about tainted breath, a concern shared even by the immortals – Cybele, an ancient mother goddess, refused those with garlic-breath entrance to her temple. Brahmins, the Hindu priestly caste, also eschewed garlic. Overly stimulating, the smell distracted from prayerful contemplation and smacked of the inferiors who consumed it.

The Romans took their culinary alliums with them into other parts of Europe to join the chives (*A. schoenoprasum*) that grew wild there. The cultivars held their place in the subsequent monastery and physic gardens. Their popularity in elite cultures – whose recipes appear much earlier in cookery books – was tempered once again by concern over the smell, and fears about the sometimes violent heating action on the digestion. Garlic, as ever, was the main culprit. Ladies, or those who courted them, should permit only the merest hint of garlic said the 17th-century diarist John Evelyn, by gently rubbing round the dish with a single clove, or rely upon the safer, gentler onion. Peasants couldn't afford to be fussy, but perhaps they had the last laugh.

Those peasant dishes of the Mediterranean diet, heavily based on garlic and onions, have turned out to be healthy as well as delicious. Alliums contain carbo-hydrates, or sugars, which provide the savoury sweetness of French onion soup or red onion marmalade. In addition to simple sugars, onions, garlic, shallots and leeks contain relatively large amounts of more complex carbohydrates (fructo-oligosaccharides). Indigestible in the small intestine (humans lack the necessary enzyme), these pass on to the colon where they are fermented by the healthy gut bacteria, which they encourage to the detriment of harmful ones, and thus function as a probiotic food. Added to their potential in defending against blood clots, help in managing diabetes and possibly preventing cancers, alliums have a great deal to offer within those pungent, encircling layers.

Opposite left and right Leek and garlic (*Allium porrum* and *A. sativum*), both from Joseph Jacob Plenck's *Icones Plantarum Medicinalium*. This Austrian doctor and dermatologist compiled seven fabulous illustrated volumes of medicinal plants (1788–92); an eighth was added after his death. The inclusion of these familiar kitchen vegetables reminds us of their long medicinal history, as well as the overlap between these categories.

Brassicas

Brassica spp.

Eat Your Greens

'The time has come,' the Walrus said, 'To talk of many things:
Of shoes and ships and sealing-wax
Of cabbages and kings...'

Lewis Carroll, 1872

The humble brassica is a large family. We exploit the leafy greens of cabbages and kales, as well as other parts in the case of broccoli, cauliflower, Brussels sprouts and kohlrabi (all varieties of *B. oleracea*), and the roots of turnips and swedes (*B. rapa* subsp. *rapa* and *B. napus* subsp. *rapifera*) – but all lack glamour. Perhaps familiarity has bred contempt. These vegetables are often associated with bitterness in taste, sulphurous odours in cooking and flatulence. Yet the Germans savour their sauerkrauts (pickled cabbage); for the Russians shchi (cabbage soup) is a national dish; and the Dutch salad coleslaw now accompanies American fast-food wherever it appears. In East Asia brassicas are also prized – the pickle kimch'i is an essential component of a Korean meal and in recent years numerous Chinese leaves and bok choi have moved with the wok into Western kitchens. Mustard, from brassica seeds (*B. nigra*), is an ancient condiment of the Mediterranean lands. Today, improved rapeseed (*B. napus* subsp. *oleifera*) is one of the world's most important oilseed crops and its residues a much-used animal feed. As well as their distinctive taste, brassicas can be defined by their bountiful diversity. Uniquely, all parts of the plant have been developed for use over time, from the age of prehistoric foraging to industrial agriculture.

Brassicas' variability, exploited many times during the history of domestication, is due to the complex history of their genome. Over the twenty million years since the early progenitor of today's brassicas appeared, a series of hybridization or gene duplication events has had the effect of multiplying the number of chromosomes, so that by five million years ago brassicas had four copies in their genome. Some four million years ago assorted species of brassica (including the edible *B. oleracea*, *B. rapa* and *B. nigra*) diverged from their common ancestor. A final natural hybridization between members of the species occurred just 2,000 years ago, when *B. napus*, the oilseed plant with its huge genome, joined the mix.

Early domestication of a brassica (*B. rapa*) may have been for the plant's oily seeds, although the leaves and swollen roots were also available. Purposeful cultivation from around 2000 BC occurred over a wide area, from the Mediterranean

Opposite Ornamental but absolutely edible varieties of kale or borecole (loose- and open-headed) from *Album Benary* (1879). The central one is a palm-tree kale, also known as cavolo nero, which has become fashionable again as an heirloom variety.

Below Fields of rapeseed oil plants colour the agricultural landscape a bright yellow. Varieties bred for consumption have low levels of erucic acid and bitter-tasting glucosinolate. Before these advances, the oil was a leading lubricant for steam engines. Suitably modified, it may make a comeback as an environmentally friendly lubricant in motor engines.

to India, when plants appeared as weeds in the land prepared for cereals. After the oil came the leaves of *B. oleracea*. Wild brassicas were native to Europe's Atlantic and Mediterranean coasts. The early cultivated plants, similar to modern kales or collard greens and lacking pronounced heads, were widely enjoyed by the Celts, Greeks, Romans and Egyptians.

Cabbage was held in high regard as a medicinal plant; even the urine of cabbage eaters, applied externally, had a place in the *materia medica* according to Cato the Elder, writing in the 2nd century BC. The Romans took their stalky garden varieties with them to lands they conquered, but densely headed cabbages, with leaves tightly wrapped around the stalk to form a heart, were developed in northern Europe. Pliny wrote about, but probably didn't eat them in the 1st century AD; they don't grow well in hot climates and were not widespread in Europe before the Middle Ages.

Inevitably the garden plants continued to interbreed with wild and feral populations as people began to fiddle around with different parts of the plants and utilize their inherent variability. Broccolis gave rise to cauliflowers perhaps in southern Italy via a Sicilian calabrese-type plant, although Cyprus is mentioned as a possible origin too. Both varieties exploit the growth of flowering tissues in the plant, holding it at the bud stage when the nutrients (which would be used for the flowers) are abundant. The fattened stem of kohlrabi was first reported in 16th-century Germany. There are hints in the historical record about sprouts from their eponymous city of Brussels in the 13th century. Perhaps the most divisive of the brassicas in the taste stakes, sprouts reputedly appeared on the menu of a 15th-century wedding feast at the Burgundian court, making the transition to Christmas dinner essential in some European countries in the 19th century, where fresh (and preserved) cabbage dishes were already popular. The role in winter celebrations served as a reminder that these greens could stand through the coldest part of the year.

Despite their prominence today, brassicas were not the most important leafy vegetables in the early husbandry of northern China. This role was filled by the perennial mallow, *Malva sylvestris*, its slippery, mucilaginous properties compensating for its lack of vegetable oils. Oil extraction techniques advanced, the mallow became forgotten, while the range of brassica cultivars that could provide food all the year round increased, and these were then spread around East Asia and crossed with local varieties.

A bitter taste often indicates the presence of a toxic substance in a plant. Our taste buds pick this up, although not nearly as well as our hominin ancestors must have done before the use of fire for cooking. It is still the bitterness of brassicas that their haters object to. Brassicas

An 'earliest solid blood red Erfurt' cabbage. Erfurt, in the Thuringian basin of central Germany, is renowned for its long history of growing vegetables.

HERB. HORT. KEW.

The Wild Flora of Kew Gardens
Name: *Brassica nigra* (L.) W.D.J.Koch
Vern. name: Black Mustard
Location: North Arboretum: outside Wing A of the Herbarium (zone 113)
Notes: Grown from spillage from a birdfeeder.
Date: 19 May 2011
Collector: T.A. Cope   No.: RBG 480

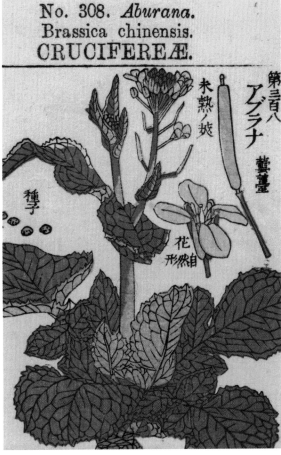

No. 308. *Aburana.*
Brassica chinensis.
CRUCIFEREÆ.

are rich in sulphur-containing glucosinolates. When the plant tissues are damaged by cutting, cooking or interaction with human gut flora, an enzyme breaks them down, and volatile sulphur is released – hence the smells – and mustard oils or isothiocynates are produced. These serve the plant by deterring predators and give the pungency. They may also be responsible for the cancer-preventing properties currently being investigated. Mothers may well have been right all along – you should eat your greens.

Above left A herbarium specimen of black mustard seed (*Brassica nigra*). Used as a condiment in Europe and Asia, its seeds were also part of an early weights system in the Indus valley.

Above right Brassica chinensis or Aburana (the Japanese name for rapeseed) from *The Useful Plants of Japan* (1895). Produced by the Agricultural Society of Japan, this volume informed its readers that as well as using the oil for cooking and lamps, the flower buds and leaves could be boiled or preserved in salt.

Asparagus

Asparagus officinalis

A Delicacy Ancient and Modern

Of all the garden plants asparagus is the one that requires the most delicate attention.

Pliny, 1st century AD

'Garden sperage' from an illustrated edition of Gerard's *Herball* (1633). The immature stems of other foraged plants, such as wild or cultivated hops (*Humulus lupulus*), were also eaten, but asparagus had long set the ideal in spring shoots.

Cultivated asparagus has been a delicacy since it was first taken into the garden. During its season this perennial vegetable requires harvesting daily by hand. Each plant has an eight-week cutting window after which it must be allowed to grow undisturbed. Establishing a bed takes three or four years, when little can be cut; it can remain productive for up to twenty years, generating green or purple spears depending on variety (white ones are blanched by earthing up). But asparagus is land hungry, likes choice soils, benefits from regular manuring and must be weeded carefully to avoid damaging the underground crowns.

The Roman author Cato gives remarkably similar advice on cultivating asparagus in his *On agriculture* (160 BC), but its labour-intensive character wouldn't have bothered him. His instructions were for estates farmed by slaves – cultivating asparagus and its consumption were luxuries. Including it in an appendix to his work, Cato was perhaps adding a relatively new introduction to the Roman garden. The beginnings of the cultivated *Asparagus officinalis* are uncertain. Probably in the eastern Mediterranean or Asia Minor, collectors of wild asparagus noticed then (as now) that returning each year to cut the spears in the spring increased the yield of individual plants. 'Asparagus' is derived from the Persian 'asparag', a general term for young shoot. Pliny rejoiced that wild plants – *A. acutifolius* – were easily found. Some considered the thinner wild spears to be superior to the fatter garden crop; the taste was certainly held to be stronger.

Garden asparagus appears to have faded after Roman decline, though where the Muslims held sway it continued to be grown. It maintained a presence in the medicinal plot too, often in monasteries. The water in which it was boiled was drunk as an aphrodisiac, but it was the seed and especially the root that were most used. Decoctions of the root were used as diuretics and this raises a vexed problem. The 16th-century Italian, Alessandro Petronio, was uneasy about the wholesomeness of asparagus. He reasoned that if asparagus made the urine smell foul, as it seemed to, it was putrefying dangerously in the body. Others would echo his reports of stinking urine, but asparagus's popularity grew. France's Louis XIV championed it at Versailles and the diarist Samuel Pepys reported eating and

Sparagus.   ⎱1. The Grass.   4. Berry⎰   Asparagus.
Eliz. Blackwell delin. sculp. et Pinx. ⎰2. Flower.   5. Seed⎰
                             ⎱3. Flower separate.

cutting what he referred to as 'sparrow grass'. The smell may be caused by the breakdown of the sulphur compound asparagusic acid, but it seems that only some people process this acid to produce the odour and only some can detect it.

Before the days of chilled airfreight, fresh asparagus travelled only short distances – the condition of the spears and their taste deteriorate quickly. Its flavour, which has tantalized over the centuries, owes part of its appeal and complexity to the fifth basic taste – umami. Long known in East Asia, it has recently been accepted more widely, along with the familiar sweet, sour, salty and bitter. In 1912 the Japanese chemist Ikeda Kikunae asked his sceptical audience to conjure what was common in the flavour of 'asparagus, tomatoes, cheese or meat'. It was umami, difficult to translate but meaning tasty or savoury. Ikeda identified umami as the amino acid glutamate; asparagus contains relatively high amounts.

'Grass', flowers, berries and seed of asparagus. One of 500 plates of medicinal plants drawn, engraved, coloured and published by Elizabeth Blackwell in her *A curious herbal* (1737–38/39). Blackwell undertook this gargantuan task to raise money to secure the release of her publisher husband from debtor's prison. She drew the plants from life at Chelsea Physic Garden, London.

Hop

Humulus lupulus

The Bitter in Beer

The hop doth live and flourish by embracing and taking hold of poles, pearches, and other things upon which it climeth … the floures hang downe by clusters … strong of smell.
John Gerard, 1636

The homeland of the common hop stretches across Europe into Central Asia as far as the Altai Mountains. A herbaceous perennial, it puts up its new shoots from extensive underground rhizomes each spring. The tender tips and young leaves were reportedly eaten at the time Pliny was writing his *Natural History* in the 1st century AD, and hop shoots are still a delicacy in parts of Europe.

Hops have a long history of medical use, relating to the bitterness found especially in the inflorescences or 'cones' of the female plants (*Humulus* are dioecious, having separate male and female plants). The cones are made up of bracts; at the base of each is a gland that produces the bitter acids humulone and lupulone. Hop medicaments were thought to cut through unhealthy, thick humours, cleansing the body and restoring flow, though hops were also thought to induce melancholy.

Quite when and how hops were first used in brewing is not known. If hops are added to the infusion of malted grain or wort, and the resulting mixture is boiled prior to fermentation, it is sterilized. The heat releases the acids from the hops, which react with proteins from the grains, clearing the brew, and their continued antimicrobial effect keeps it from taint. It was for this reason that small quantities of hops were eventually added to all commercial beers, even when a heavily hopped taste was not necessarily desired.

The keeping qualities of hopped beer meant that brewing could be turned from a local business into a viable trade. By the 13th century Bremen in Germany had a lively export business with Flanders and the Netherlands. Imports of hopped beer in 14th-century England were mostly for people of German and Dutch descent, until some started brewing themselves. King Henry VIII didn't like hopped beer, but later Tudor monarchs encouraged hop production. Their aim was the victualling of the army and navy: hopped beer provided a clean source of portable refreshment that would keep. And late in 1620 as conditions worsened on the *Mayflower* after its Atlantic crossing, shortage of beer was reportedly one reason the Pilgrims decided to make their settlement at New Plymouth rather than sail on.

North America had its own variety of hop (*Humulus lupulus* var. *lupuloides*), but the European plant made its way over in about 1630. This export was followed

HUMULUS LUPULUS.
Der gemeine Hopfen.

The habitat of the common hop plant was the wetlands populated by alders and oaks. After the climatic and human changes of around 6000 BP, the hop flourished at the edges of the reducing woodlands, in boggy bottoms and opportunistically clothed hedges, making it easy to forage from the wild.

by the familiar attempt to grow key crops as part of imperial and settler projects. The original India Pale Ale, exported from Britain from the 18th century onwards, gained its special taste from the long sea voyage and further maturation in tropical temperatures. In the late 19th century entrepreneurs at the Murree Brewery in Rawalpindi, and would-be hop growers in Kashmir and Himachal Pradesh, all contacted Kew Gardens for help with their plants.

The right amount of beer has a soporific effect, but if you prefer not to take your hops in this form, a hop pillow is reputed to do the same.

Tomato

Solanum lycopersicum

The Love Apple

Solanum peruvianum from *Curtis's Botanical Magazine* (1828). The 'Large-flowered Tomato' is a native of Peru and Chile. Although it will not naturally hybridize with *S. lycopersicum*, genetic techniques allow researchers to exploit the disease- and parasite-resistance of this wild tomato relative.

How did the Italians eat spaghetti before the advent of the tomato? Was there such a thing as tomato-less Neapolitan pizza?

Elizabeth David, 1984

The tomato, one of several fruits treated as if it were a vegetable, originated in western South America. Perhaps regarded by the peoples of the Andes as an uninteresting weed, once its seeds were spread, at least partially by birds, it was domesticated in Mesoamerica. The original plant had small cherry-type fruit, but by the time of the Spanish conquest of the early 16th century tomatoes in a range of sizes, colours and textures were available in Aztec markets. The Aztecs ate them raw and combined them with chillies to make spicy sauces. 'Tomato' is derived from an Aztec word, although early descriptions sometimes confused the tomato with a variety of physalis, a native plant with small green fruits that the Aztecs also ate.

The Spanish tried them, liked what they tasted and returned to Europe with seeds. Tomatoes were not immediately popular, however; for one thing, the leaves, which are inedible, resemble those of the poisonous deadly nightshade (*Atropa belladonna*), another member of the same family (Solanaceae), as is the potato. One early naturalist grouped the tomato with the mandrake, perhaps because of a resemblance of their roots. It thus acquired a brief reputation as an aphrodisiac.

The tomato was more favourably received in Italy, where it gradually became a mainstay in cooking. It slowly spread throughout the Mediterranean countries and further north, where plants were sometimes grown as ornamentals. The Turks took tomatoes to the Levant and the Balkan countries. By the 19th century commercial production was underway in Italy, and the tomato was introduced to the United States. Tomatoes became the main ingredient in 'ketchup', a Chinese-derived word that initially referred to a spicy fish sauce. 'Tomato ketchup' became so dominant that it is the default version for such condiments.

Despite being a slow starter, the tomato has now become a worldwide favourite. In China, where it was introduced via the Philippines in the 16th century, it was not grown much until the 20th century. Today China is the world's largest producer. The British introduced the tomato to India, now the second largest producer, in the late 18th century. Its initial modest cultivation there was primarily for European palates, but tomato-based dishes have become integral to Indian cuisine. In fact, the tomato is second only to the potato in worldwide production.

The tomato's commercial importance has led to genetic research to improve yields and pest resistance, produce fruits with a rich colour and increase the thickness of the skin for ease of transport and longer keeping. Unfortunately, the flavour generally suffers, and supermarket tomatoes, available all year round, lack the taste of a good fruit. The distinctive flavour and smell of a ripe tomato come from a complex series of chemicals, including volatile aromatics, acids and sugars, some of which have been partially bred out of commercial varieties. Scientists are trying to breed them back in, although the experiment with genetically modified tomatoes in the 1990s has been abandoned. Commercial tomatoes are often picked green for transportation, and ethylene, produced naturally in the ripening process, is sprayed on them when ready for display. These issues have encouraged increased interest in 'heritage' (or 'heirloom') tomatoes, many of them older varieties.

A selection of 'Tomatoes or Love-Apples' from *Album Benary* (1879). Many of these are now considered heritage or heirloom varieties and are not commercially grown. All live up to the name 'tomatl' bestowed on them in the Mexican Nahua peoples, which means 'plants bearing globous and juicy fruit'.

Heal and Harm
Getting the Balance Right

Scarified seed heads of the opium poppy, *Papaver somniferum*, with traces of the dried juice that is processed to produce the drug.

Throughout recorded human history, and undoubtedly before, plants have provided the basis of therapeutics, and there are few plants in this volume that have not been used by someone, somewhere, at some time, to try to alleviate the ills or injuries that plagued them. But the plants in this section possess compounds with specific physiological effects on the human body, and some of them are still important in modern scientific medicine.

Anything, in certain circumstances, can be a poison, and the balance between heal and harm is always finely judged. Cocaine, the alkaloid from the South American plant *Erythroxylum coca*, is a powerful local anaesthetic, but also central to the lucrative international trade in illicit drugs. Opium, an effective remedy for pain, can be extracted from the poppy, which is the source of morphine too, another useful drug, but also of heroin, ironically introduced as a less addictive alternative to morphine.

The willow tree has a long use in medicine, and its main therapeutic ingredient, slightly modified, yielded aspirin. Preparations of aloe can also be found in the home medicine cabinet, and it is an ingredient of many cosmetics. Rhubarb is now mostly enjoyed as an early spring fruit, instead of as a purgative, which its roots can produce. Citrus fruits, a relatively late arrival in Europe in any quantity, were appreciated for

their power to both prevent and cure scurvy, a common disease of long voyages and among land people with limited winter diets.

Two species of *Strychnos,* one from Asia and one from South America, produce substances which swing the balance between healing and harm towards the latter. This did not prevent *S. nux-vomica,* the Asian species, from enjoying a long history as a tonic. Its active ingredient, strychnine, was also used as a rat poison and an agent of murder. The South American counterpart yielded curare, which kills by paralysing muscle action. It found use in surgery before better relaxing agents were discovered. Ancient Indian doctors used the roots of *Rauvolfia* to treat snake bites and much else. Its active alkaloid, reserpine, enjoyed a brief period of use in the West, when it was found to lower blood pressure. It also seemed promising in psychiatric disorders, but, as with curare, other drugs soon replaced it. On the other hand, both quinine, from South America, and artemisinin, from China, are still important in the battle against malaria, a major modern killer.

Plants can also produce steroids, and the Mexican yam was the source of the cheap steroids that yielded the contraceptive pill ('the Pill'), one of the most important medical innovations of the past half-century. It needed modern chemistry to modify it, and chemistry has also played its part in turning a substance from the Madagascar periwinkle into a treatment for childhood leukaemia.

Above left Strychnos nux-vomica drawn by one of the Indian artists employed by the surgeon-botanist William Roxburgh to illustrate the plants he had collected from Coromandel, the east coast of Madras in the 1780s and 1790s.

Above right One of the many distinctive aloes of the southern Africa 'hotspot', described before botanists used the term but were aware of the variety of species, as featured in G. K. Knorr, *Thesaurus rei herbariæ hortensisque universalis* (1770–72).

Poppy
Papaver somniferum
Pleasure, Pain and Addiction

All [Mrs. —'s] friends advised her to lay aside the use of opium, lest it should by habit become necessary; but she whispered me privately, that she would rather lay aside her friends.

George Young, 1753

The field poppy (*Papaver rhoeas*) is native to Europe, where it grows almost every-where, mostly as a pretty weed of agriculture; it is often red. The opium poppy (*P. somniferum*), generally white, is also vigorous and has long been deliberately grown for its main product: opium. The poppy was in fact among the earliest cultivated plants, probably first tamed in the western Mediterranean, known to Neolithic man and appreciated by early Mediterranean cultures, including in Egypt, Crete, Greece and Rome, and used as far east as India. The Hippocratics (5th–4th century BC) recommended it for pain control as well as for treating diar-rhoea and many other diseases, or for producing sleep. Dioscorides, the great 1st-century AD authority on drugs, described opium as having cooling properties, thereby useful for hot conditions.

Harvesting the poppy is not easy: the unripe capsule must be scored at the correct time to exude the crude opium juice, which is collected and processed by drying in the sun and boiling. What was originally a sticky white liquid becomes a brown paste, and further sun drying produces a brown, clay-like substance, much richer in opium and easy to mould into cakes or rounds for transport. Knowledge of these techniques is prehistoric: small juglets made in Cyprus and exported to Egypt and elsewhere, which contained opium, resemble the inverted shape of the poppy head and are often scored or painted in imitation of incisions.

That ingesting opium could produce a sense of euphoria was long known; so was the fact that ever-larger doses were often necessary to achieve the same effect, as well as the substance's ability to create dependence in its users. An overdose could kill, and opium was undoubtedly used in murder, including reportedly by the emperor Nero (37–68). A later, more admired emperor, Marcus Aurelius (121–80), whose personal physician was Galen, used opium regularly, but was apparently able to control his dosage and thereby maintain self-control. The 'bondage of opium' over the centuries has been highly variable in different regular users.

Most opium habitués probably first encountered the substance medicinally, as it was a mainstay in the doctor's armamentarium – 'God's own medicine', accord-ing to Sir William Osler, the 19th-century Canadian physician – and used at times

Opposite Poppies had become popular by the 16th century, manipulated into different colours and shapes from the native plants of Asia. Turkey was a major source of garden specimens. This watercolour of a red-fringed opium poppy by the German Sebastian Schedel appeared in his *Calendarium* (1610), which was arranged month by month as the flowers came into bloom.

Above The 'Pavot' (poppy) from J. J. Grandville's *Fleurs animées* (1847) sprinkles her seeds and puts the insects to sleep. The accompanying poem 'Nocturne' (by Taxile Delord) refers not just to peaceful sleep but also to the power of narcotic-induced dreams. The Romantic poets Coleridge and De Quincey both reported using opium to heighten their creative powers.

for virtually every disease or symptom. 'Opium and lies' was Osler's pungent recommendation for treating advanced tuberculosis. The reforming 16th-century doctor and alchemist Paracelsus might have rejected much of traditional medical knowledge, but he retained opium as a favourite remedy. He coined the term 'laudanum', although it was the recipe of Thomas Sydenham in the 17th century that became standard: a mixture of opium and red wine, spiced with saffron, cloves and cinnamon. So enthusiastic was Sydenham about opium that he was dubbed 'Opiophilos'.

During the 18th and 19th centuries, many patent medicines, including Dover's Powders, Godfrey's Cordial and Daffy's Elixir, contained opium (and generally alcohol). Such medicines were used as cure-alls, as well as for keeping children quiet and soothing frayed nerves. Their unregulated sale until well into the 19th century meant that they provided a major source of income for quacks, druggists and apothecaries.

The addictive properties of opium led to social and legal concerns, even if many men of letters, most famously the poet Thomas De Quincey, statesmen and others of all classes, occupations and both sexes depended on its daily physical and psychological effects. Governments, too, valued the revenue its import raised, and this was especially true in British India. India and Turkey were the two primary sources, with Turkish opium generally used in Europe. However, the opium poppy was also widely cultivated in India, both for local use and, increasingly, for export to China, which since the 18th century had provided a market for social use (it had earlier been employed there medicinally). There was some official disquiet in Britain that opium export to China was not entirely ethical, as it created a culture of dependency, and Chinese authorities legislated against its importation and use in the 19th century. The two Opium Wars (1839–42 and 1858–60) were about much more than just opium, but they did forcibly smooth the Indian-Chinese trade.

Poppy Seed Head, by Brigid Edwards. The dried seed capsule functions like a pepper pot, sprinkling the seeds from the pores. These natural structures provided the inspiration for ancient jewelry and pottery. Small jugs were exported from Cyprus in the 2nd millennium BC to Egypt, Syria and Palestine. When inverted, their shape mimics the poppy capsule and enough of their original contents has survived to reveal that they were used to store opium.

In the meantime, the chemical contents of the opium juice were analysed. Its most potent alkaloid, morphine, named after the Greek god of sleep, was isolated in 1804 and marketed a couple of decades later. Codeine, currently the most widely used opium derivative, followed in 1832, and these discoveries allowed pharmaceutical entrepreneurs to market the products in more or less pure form. Heroin, chemically adapted from morphine, was introduced in 1898 as a 'safer' alternative to the latter. The development of syringes in the 1850s made the effects of all opiates (and other substances such as cocaine, another plant-derived substance) much more powerful, and increased addiction and dependency. National and then international attempts to control the sale and use of opiates and other addictive substances followed.

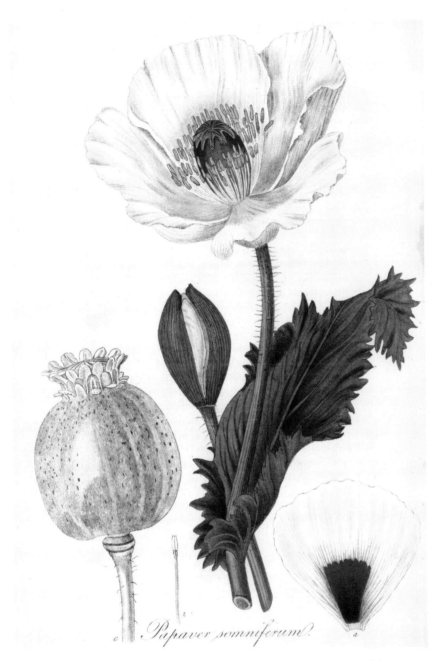

Papaver somniferum

The 'war on drugs' from the early 20th century has not been a notable success. Making them either prescription-only, or completely illegal, has had the unfortunate consequence of driving supply and use underground, and created a criminal class to exploit the market, estimated to be worth some $350 billion a year. The rise of AIDS in the 1980s compounded the health issues, as the use of dirty needles spread this and other diseases among users. Afghanistan has emerged as the world's leading supplier of opium, but the Golden Triangle countries of Southeast Asia, as well as Colombia, are also leading players. Morphine and codeine are still widely used in medicine, but the delicate balance between heal and harm remains.

The classic white opium poppy. The Greek doctor Galen (129–c. 210) extolled the virtues of Olympic Victor's Dark Ointment. This salve contained opium and dried to an elastic patch. It provided an external way to relieve pain and swelling, especially around the eyes.

Tab. 131.

CINCHONA OFFICINALIS L
Die gemeine Fieberrinde.

Cinchona, Artemisia

Cinchona officinalis, Artemisia annua

Fighting Malaria

The Peruvian bark became my sheet anchor.

Thomas Sydenham, 1680

For more than two centuries, 'the Bark' was shorthand for a medicine derived from *Cinchona*, a genus of about forty species of evergreen trees indigenous to the elevated slopes of the South American Andes. Two species, *C. officinalis* and *C. pubescens*, and a cultivar, 'Ledgeriana', contain significant amounts of alkaloids, of which quinine and quinidine are especially important.

To the indigenous South American peoples 'quinquina' meant 'bark of barks', and cinchona bark was undoubtedly used medicinally in pre-Columbian Peru and other Andean areas. Malaria probably arrived only with Europeans, although the mosquitoes that we now know transmit the disease were already there. The story that the bark was recommended to treat the malarial fever of the Countess of Chinchon in 1638 is apocryphal, but accepted by Linnaeus, who misspelled her name when creating the genus *Cinchona*. Spanish physicians and missionaries learnt of the value of *Cinchona* bark in treating intermittent fevers (generally what we would call malaria), and shipped the new drug back to Europe, where malaria was common.

Variously called Peruvian Bark, Jesuit's Bark or simply the Bark, the new remedy for fever established itself in the 17th century when an enterprising English practitioner used it at the French court. At that time the diagnosis and treatment of 'fever' were based simply on symptoms or individual clinical experience, and the bark varied in quality. Nevertheless, it was a significant source of income for Spanish Peru, which meant that access to the trees was carefully controlled.

Two French chemists, Joseph Bienaimé Caventou and Pierre Joseph Pelletier, isolated the active principle, quinine, in the early 19th century. This made dosage much easier to manage, and with European imperial expansion in Africa and Asia demand was very high. Both the British and the Dutch were keen to smuggle seeds out of Peru. Behind the British initiative was Joseph Dalton Hooker, director of Kew Gardens from 1865 to 1885,

PURE QUININE.
5 grains one dose.
Price 1 pice.

whose extensive experiences in India, where malaria was rife, allowed him to see the potential for establishing *Cinchona* plantations there. The Dutch eyed their possessions in Java.

There were farcical elements in these early attempts at botanical espionage, as seeds from the wrong species were obtained and the long voyage back to England took its toll. By 1861 Kew had seeds, and seedlings destined for India were raised. These were transported in Wardian cases, special sealed glass boxes used in shipping plants in the 19th century. Plantations were established in India, Ceylon (Sri Lanka) and Java, easing the supply of quinine, although the need was always great. The bark was gathered in the traditional way by cutting from only part of the tree so that it could recover, and then drying the product in the sun before processing.

In the late 1890s, Ronald Ross and Giovanni Battista Grassi discovered the life cycle of the *Plasmodium* parasite that causes malaria and its method of transmission by the bite of female *Anopheles* mosquitoes. This helped explain its geographical prevalence in marshy areas and tropical lands where frequent thunderstorms left breeding puddles. Mosquito control became part of the preventative strategy, but quinine, taken as a prophylactic and as treatment, remained essential. Disruption of supplies during the Second World War encouraged the development of alternative synthetic drugs, of which there are now many. Although quinine-resistant strains of *Plasmodium* now exist, quinine long remained effective and is still used in treating some forms of the disease; quinidine, the other main alkaloid in *Cinchona*, is used to treat heartbeat irregularities.

The causative parasite of malaria has unfortunately acquired resistance to all front-line treatments, but in the battle against the disease the plant world has yielded another drug for the modern armamentarium. Sweet wormwood (*Artemisia annua*) is a bit of a weed. It grows easily on slopes, at the forest limits and on waste ground disturbed by human activity. It also contains a powerful antimalarial compound, artemisinin (*qinghaosu*). This and its derivatives, which allow the drug to be given by injection as well as orally, stepped into the breach as drug-resistant malaria reached frightening levels in Southeast Asia in the aftermath of the region's wars and population displacements. Malaria, much like wormwood, thrives in chaos.

Qinghao (*A. annua*) had been part of the Chinese pharmacopeia since the 2nd century BC. What brought it to the fore internationally was a request from North Vietnam during the Vietnam War for China's help with anti-malarial drugs and the circumstances in China at the time. Mao sought to both destroy and utilize China's past; modern science would strip the best from traditional medical texts and make China self-sufficient in medicinal products. In 1967 a secret project – 523 – was set up to screen plants such as *A. annua* for their anti-malarial potential. The pharmacologist Tu Youyou led the team, and as well as the source plants,

she also paid attention to the original method of producing the drug, as described in the 4th century AD by Ge Hong in his *Emergency Prescriptions Kept Up One's Sleeve*. He was the first to recommend *qinghao* for 'intermittent fevers' and his instructions provided a vital clue that great heat should not be used in the extraction process. In isolating the active component of the plant extract, Professor Tu produced the first water-soluble artemisinin derivative.

Although there was some scepticism when the rest of the world learnt of China's new anti-malarial drug in 1979, today various artemisinin drugs are a crucial part of the World Health Organization recommended ACT (artemisinin combination therapy), which uses a cocktail approach to curtail parasite resistance. The cat-and-mouse battle between drug and parasite continues.

The Chinese anti-malarial *Artemisia annua* from *Somoku-Dzusetsu; or, an iconography of plants indigenous to, cultivated in, or introduced into Japan* (1874). The second edition was edited by two botanists, Tanaka Yoshio and Ono Motoyoshi, who were involved with Japan's modernization. Tanaka is known as the 'father of museums' and Ono edited a bilingual dictionary of botany, much used in China.

Rauvolfia

Rauvolfia serpentina

Ancient Ayurvedic Drug

Only in the last twenty-five years has the scientific world begun to discover the true value of Rauvolfia. Credit must be given to the chemists and pharmacologists in India who started to analyse crude extracts of R. serpentina.

Jurg A. Schneider, 1955

Rauvolfia (often spelled 'rauwolfia') is a large genus of about 200 species of trees and shrubs with a wide distribution in tropical climates. It is frequently considered as a weed, so vigorous is its growth. The genus was named for the 16th-century German doctor and naturalist Leonhard Rauwolf, who travelled widely in Southwest Asia and described many new species of plants and animals, though he never actually saw any examples of 'his' genus, which was named later in his honour.

In India, where it is known as Sarpagandha, Chandra and Chotachand, the dried roots and leaves of *Rauvolfia serpentina* have long had a place in Ayurvedic medicine. Powder made from the plant had a variety of uses, including as an anti-dote against snakebite. It was also employed to calm agitated patients, and to treat insect stings and diarrhoea. In the West rauvolfia was regarded as little more than a curiosity. Then, in 1931, two Indian scientists analysed the ground root powder and identified several alkaloids. It was assumed that the root's physiological effects came from a cocktail of the alkaloids, but many laboratories, both in India and elsewhere, began to study the chemical makeup in detail and the most potent of these many alkaloids, called reserpine, was isolated in 1952.

The whole root had already been shown to lower blood pressure in laboratory animals, and this newly isolated alkaloid came at a time when high blood pressure had been implicated in rising rates of heart disease and stroke throughout the Western world. In the 1950s there were few safe drugs that could reduce blood pressure and major surgery was sometimes resorted to, involving cutting along the spine and thus disrupting the sympathetic nerve fibres, which act to reduce arterial diameter. It was dangerous and left the patient with serious side effects.

The promise of this new drug led to a systematic survey and chemical analysis of rauvolfia plants from around the world. Many were found to contain reserpine, and supplies of the drug came from Pakistan, Sri Lanka, Burma (Myanmar) and Thailand as well as India. This was important, since the Indian authorities had been reluctant to give up their hold on the new wonder drug. From the late 1950s, after it was approved for patient use in the US and Britain, the main use of reserpine was to treat raised blood pressure. However, it also created much excitement

Watercolour of *Rauvolfia*, believed to be from the collection of Claude Martin (1735–1800). Born in France, Martin deserted his native army and thrived in that of the British East India Company, settling in Lucknow. Here he established a museum and commissioned Indian artists to paint the plants and birds of his adopted country.

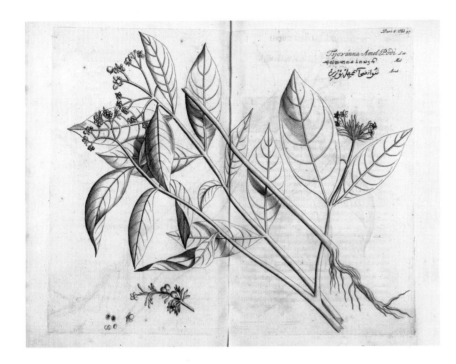

Rauvolfia, from Hendrik van Rheede's *Hortus Malabaricus*. It is not surprising that traditional healers in the Indian subcontinent tried to find remedies against snakebite. Recent research finds that the region bears the greatest burden of envenoming and death in the world, with 81,000 venomous bites and nearly 11,000 deaths each year.

in psychiatric circles, since the drug had a calming effect too. It began to be given to patients in psychiatric hospitals, where improvement was reported in schizophrenics whose behaviour became more tractable.

The early promise did not hold. Reserpine's reported side effects included depression, with several suicides attributed to its use. Intense pharmacological research into how the drug worked suggested that it might actually relieve depression rather than cause it. It acts on the nerve endings in the nervous system, blocking the release of several chemicals (called monoamines) crucially involved in activity in the central nervous system. This helps explain its calming influence, but also fed into an early theory about a biological cause of depression. There are now far better drugs available for both blood pressure control and mental disorders, so reserpine has retreated into the laboratory, where it continues to be used by scientists seeking to understand the minute workings of the brain.

Coca

Erythroxylum coca

Stimulant and Nerve Blocker

For me, there still remains the cocaine bottle.

Sherlock Holmes, in *The Sign of Four*, Arthur Conan Doyle, 1890

For the Incas coca was sacred; it was so important that the state monopolized its production and distribution, and coca leaves were offered to the gods. A small tree or shrub, coca grows best on the lower slopes of the tropical Andes and its use in the area has a long history. The leaves, which can be harvested several times a year, contain a potent cocktail of alkaloids, of which the most significant is cocaine; they also contain a small amount of caffeine.

When gathered, dried and mixed with some lime, the leaves can be placed between the cheeks and gums, where their alkaloids are slowly absorbed. The result is an increase in muscle strength, a reduction of hunger and a general sense of alertness. Spanish conquistadors described the plant and its effects, and when they forced native peoples to labour in their mines, they ensured that they were supplied with the leaves to increase production. In the mid-19th century two German chemists independently isolated its most potent alkaloid. One of them, Albert Niemann, gave it the name that has stuck: cocaine.

Cocaine was one of many plant alkaloids attracting medical and chemical attention at the time; the young Sigmund Freud, then a budding neurologist, started self-experimenting with it in 1884. His collection of papers on the subject, 'Über Coca' (On Coca), more advocacy than sober scientific analysis, came back to haunt him. He downplayed any addictive qualities of the drug, emphasizing the psychological euphoria and increase of energy and muscular strength. One of Freud's colleagues, the ophthalmologist Carl Koller, noticed at the same time that cocaine was a powerful local anaesthetic.

It took some time before cocaine's potent addictive qualities were appreciated, and in the meantime enterprising pharmaceutical companies sold the drug with a syringe for ease of administration, and it became an ingredient of popular beverages. Cocaine was also touted as a safe way to break morphine addiction. Sherlock Holmes was habituated to cocaine, although his creator, Arthur Conan Doyle, eventually broke his character's habit, as the dangers of cocaine use became more obvious. Among

Snakes and coca leaves were both sacred to the Incas. Recent analyses of Inca mummies found in a shrine in 1999 reveal that in the 12-month period leading up to their sacrificial death, as part of *capacocha* rituals, the three children ingested increasing amounts of coca leaf and *chicha* beer. This is in addition to the sizeable quid of coca leaves found clenched between the teeth of the 13-year-old girl.

Herbarium specimen of *Erythroxylum coca*. Coca leaves are browsed by the vicuña and guanaco; the alkaloids in the leaves have a local anaesthetic effect on the animals' stomachs, creating a false sense of satiety. They lose their appetite and leave the plant alone, and once the cocaine and other chemicals reach the brain, the effect of heightened energy and mild euphoria encourages them to move on. Just as these animals are not addicted to coca leaves, human use in a similar manner does not lead to the dependency associated with processed cocaine.

prominent individuals trapped by cocaine in those early days was the pioneer of aseptic surgery, William Stewart Halsted, professor of surgery at Johns Hopkins University, Baltimore. Although he managed to keep his career going, he never completely lost his addiction to morphine, which he substituted for cocaine.

Despite medical uses, cocaine has become a controlled drug in most places. Criminalizing its use has not met with marked success, and cocaine, both in its pure state and in the adulterated form of 'crack', is used by millions of people. Supplying it is highly profitable, especially for Colombia, despite encouragement for growers there to plant coffee trees instead, or for Peruvians to grow asparagus.

Strychnos
Strychnos nux-vomica, S. toxifera
Medicine as Poison

Strychnos nux-vomica, showing the smooth-shelled orange fruits, which reach the size of a large apple and contain a soft, jelly-like pulp. The seeds were reportedly used in southern India to add to the potency of distilled spirits. The wood is hard and durable, and along with the roots, this exceedingly bitter-tasting material was traditionally used to treat fevers (no doubt including malaria) and snakebite.

'Strychnine is a grand tonic, Kemp, to take the flabbiness out of a man.'
'It's the devil,' said Kemp. 'It's the palaeolithic in a bottle.'

H. G. Wells, *The Invisible Man,* 1897

In plants, alkaloids have a variety of functions including protection from animal predators, since many of them are poisonous. Two species of *Strychnos,* a genus of almost 200 trees and vines scattered throughout the tropical world, manufacture especially potent alkaloids. The main active principle of the nut of *S. nux-vomica,* from South and Southeast Asia, is strychnine, one of the first alkaloids isolated by modern chemists in the early 19th century. The South American *S. toxifera* contains curare. Each of these two alkaloids played a significant role in their indigenous cultures, as well as in the West.

Although *nux-vomica* is sometimes mistakenly assumed to mean 'emetic nut' (the *vomica* means 'depression' or 'cavity'), strychnine in fact makes it harder to vomit, as doctors trying to treat overdoses learnt to their cost. Strychnine is a nerve stimulant, producing spasms and, in high doses, convulsions and death. The nuts of the tree acquired their place in medicine because of their obvious stimulating effect and were also thought to protect against snakebite. *S. nux-vomica* has a long history in Indian Ayurvedic medicine and Arab doctors knew of the medicine in the Middle Ages. The Portuguese in Goa studied it and began importing it to Europe.

In Western medicine *nux-vomica* was prescribed for many complaints and was available over the counter for people wishing to treat themselves. With its obvious effects of tightening muscles and generally stimulating those who took it, it seemed to be a 'pick-me-up'. Even after its active principle, strychnine, was isolated the crude drug, ground from the nut, continued to be favoured. Ironically, it was also used as a rat poison as early as the 16th century, and both nut and its alkaloid could be used for more sinister purposes. Strychnine was harder to detect than the other favourite Victorian poison, arsenic, which meant that doctors, who had easy access to it, preferred it. The infamous poisoner William Palmer used it to murder his mother-in-law, (possibly) his wife and several children and a friend, and Thomas Cream, another doctor, used it serially. Both were hanged. As chemical detection methods improved, and, eventually, supplies were more systematically controlled, strychnine fell from favour. Even its use as

a poison for vermin was finally outlawed, and it faded from the scene, except in detective fiction.

While strychnine the medicine became classed as a poison, curare the poison acquired a legitimate place in medicine. In the tropical regions of South America, hunters dipped their arrows in a substance obtained from *Strychnos toxifera* as they were aware that it caused paralysis, making killing game easier. It was also used in warfare. Early European explorers naturally became intrigued, and samples of the miraculous substance reached Europe by the late 16th century. But it remained largely a curiosity until the famous French physiologist Claude Bernard showed in 1856 that curare, as the active principle was called, acted to block the junction between the motor nerves and the muscles. Death was caused by suffocation, as the muscles controlling breathing were paralysed. An experimental animal could be kept alive by artificial ventilation, and because the other muscles were also paralysed, and therefore flaccid, operations were easier to perform. The drug thus became an important adjunct to modern surgery, with ventilation controlled by the anaesthetist. Although curare has been replaced by newer relaxing drugs, working out its mode of action greatly contributed to our understanding of how nerves and muscles interact.

Two gourds from Guyana containing curare made from *Strychnos toxifera* bark and used to tip poison arrows. These were collected by Robert Spruce (1817–93), who travelled extensively in South America at the behest of Sir William Hooker at Kew and George Bentham, who described many of the specimens Spruce collected.

Rhubarb

Rheum spp.

Potent Purge to 'Superfood'

Nor was it till I had turned back the curious bracteal leaves and examined the flowers that I was persuaded of its being a true Rhubarb.

Joseph Dalton Hooker, 1855

Rhubarb has now earned the plaudit of 'superfood', a fairly meteoric rise for a plant that has only been eaten seriously for about two hundred years. As a medicinal, rhubarb is much older, but rather than the stems of the dessert vegetable it is the dark-skinned roots that are used. Rhubarb root, especially of species identified as *Rheum officinale*, *R. palmatum* and *R. tanguticum*, featured in early Chinese medical texts and was well known to folk practitioners, particularly as a cathartic.

Most of the sixty or so species of rhubarb are found on the mountainsides and desert areas of northern and Central Asia, where they have evolved into wonderfully diverse forms. Traded along the Silk Road, rhubarb root entered the Western pharmacopeias in the 1st century AD. Dioscorides described it as a stomach tonic, stimulating the appetite and assisting digestion. Persian practitioners recommended it as a stomach strengthener to counteract headaches after too much wine. It was not until the Middle Ages that its purgative properties garnered attention. As rhubarb's gentle laxative nature became better known, together with a growing awareness of the sale of many inferior roots, Europe became increasingly interested in sourcing 'true rhubarb'. The Portuguese were delighted to bring it home from Macao after opening their Chinese trading port there in the 16th century. The Dutch and British entered the competition, but the interior of China and the finest root still remained unreachably remote.

The Russians were the most successful in bringing the best rhubarb through to Amsterdam in the 17th and 18th centuries. Keen to maximize their profits, the Russian state instituted strict quality control at Kyakhta on the border with Mongolia. By this time rhubarb seeds had also made their way West, and gardeners as well those in charge of the physic and botanic gardens were trying to produce plants of medicinal quality. The Royal Society of Arts in Britain offered medals for sizeable plantations, but debate continued over which variety was the 'true' rhubarb as the potency of homegrown roots proved disappointing.

Rhubarbs do not come true from seed and readily hybridize to produce new varieties, which can be perpetuated by root division. This is perhaps what led to the development of culinary cultivars. First sold with limited success in Covent

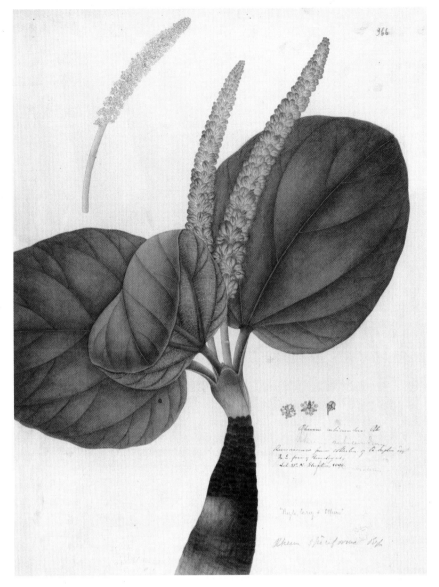

366

Garden in London in the early 19th century, the pink stalks gained in popularity after the arrival of cheap sugar sweetened their tartness. Forcing plants with warmth and by excluding light also produced a more refined delicacy than field-grown plants. The tender stems, ready early in the year, brightened the winter tables of the wealthy. In America, master plant breeder Luther Burbank produced superb new varieties.

After the Second World War rhubarb had to compete with a growing range of imported fruit and fell from favour. The resurgence in rhubarb's popularity has been driven by its new image, laden as it is with bioactive polyphenols, which are currently being investigated for anti-cancer, anti-inflammatory and other potential benefits. Like all good things, too much rhubarb over too long a period, as a purgative rather than a food, can cause problems. And don't eat the leaves.

Rheum spiciforme, like most of the 60 species of rhubarb, is native to the countries of and fringing the Qinghai–Tibetan Plateau, 'the roof of the world'. It is thought that the rapid uplift of this region led to the diversification of the rhubarb genus. The low-growing habit of *R. spiciforme* may be an adaptation to the damaging winds blowing at the great altitudes where it is found.

Willow

Salix spp.

Tree of Sorrow and Pain Reliever

There is a willow grows aslant a brook
That shows his hoar leaves in the glassy stream.
Shakespeare, *Hamlet*, Act 4, Scene 7

Few trees have been more intertwined with human history than the willow. Its use in weaving baskets and making fences and hurdles is ancient, and continues as a reviving craft. The *Salix* genus of about 300 species, with numerous cultivars, is widely spread in the temperate regions of the northern hemisphere, though willows can also be found south of the Equator and in the colder reaches of the Arctic. They frequent river-edges, where their roots help prevent erosion.

The medicinal qualities of one species in particular, *Salix alba* – the name *alba* relates to the silvery-white underside of the leaves – have long been exploited by humans. The Egyptians, ancient Greeks and peoples of Southwest Asia all used it to treat fevers and pains, and inevitably many other complaints as well. The bark was generally ground and put into an infusion of wine or other liquid. It remained a mainstay in European therapeutics, and the 17th-century English herbalist Nicholas Culpeper recommended willow bark as a substitute for the more expensive Peruvian Bark. Both have a bitter taste, and *Salix* might have alleviated the fever, though it would not have cured the disease if it happened to be malaria. In 1763, an English clergyman, Edmund Stone, specifically singled out infusions of willow bark as a remedy for fever.

It continued to attract medical attention and in the 1820s two pharmacists, one French, Henri Leroux, and the other Italian, Raffaele Piria, independently isolated the active ingredient, salicin. Further work showed that when this is broken down in the body, one of the products is salicylic acid, demonstrated in the 1870s to be valuable in reducing the pain and inflammation of rheumatic heart disease. These were the early days of germ theory and the substance was thought to be an internal disinfectant, since salicylic acid can kill bacteria in the laboratory. It is now known that it works through other mechanisms. Another species, *S. purpurea*, also contains the essential compounds.

'Aspirin', which has an acetyl molecule added to the salicylic acid, was actually synthesized in 1853, but was only marketed (with great success) by the German pharmaceutical company Bayer in 1899. It had a long life as the frontline drug for the reduction of fever, headache and pain. Its manufacture no longer needs the

willow bark; it is made from coal tar derivatives. Despite its place in the household medicine cabinet, aspirin would almost certainly have failed modern safety regulations governing the introduction of new medicines.

More recent uses of willows include as biomass, and they are coppiced for fuel in Sweden and elsewhere. Many are very fast growing and strike easily from a fresh-cut branch. Their wood is used for charcoal and making paper, their bark for tanning; they provide excellent windbreaks and their environmental impact is widely appreciated. The weeping willow (*S. babylonica*), introduced to Europe from China, is a favourite ornamental. And while *S. alba* may not enjoy its former medicinal status, another variety of the species, *caerulea*, is used to make cricket bats. No other wood will do, and the sound of willow bat striking the leather of the cricket ball is a distinctive sign of summer wherever cricket is played.

Samuel Copland extolled the virtues of *Salix alba* in *Agriculture Ancient and Modern* (1866). In addition to the familiar uses of coppiced wood as stakes and poles and for hurdles and baskets, he reported that people living near the Arctic dried the inner bark, then ground and mixed it with oatmeal as a flour in times of scarcity. In Russia it was reputedly planted and pollarded to mark the way for travellers across the Steppes.

Fig. 1

Citrus
Citrus spp.
Vitamins and Zest

Agridulce como la naranja es el sabor de la vida.
[Sour and sweet like the orange is the taste of life.]

Spanish Proverb

The oranges, grapefruits, clementines and lemons that we eat for breakfast, drink as juice or cook with may seem very familiar, yet several are of recent origin. Grapefruit (*C. paradisi*) was bred only in the 18th century, and clementines, a hybrid of the tangerine and an ornamental variety of the bitter orange, are a century younger. This is because the species of the genus *Citrus* are wonderfully fertile with each other, and the trees or bushes themselves are highly adaptable to human manipulation.

All citrus fruits originated in an area stretching from eastern Asia to Australia at a time (perhaps 20 million years ago) when Australia was still joined to the Asian continent. They have complicated historical genetic relationships, still not entirely understood, although the Australian lime (*Citrus australis*) is very old, and other citrus species have an extensive history in Asia, where they were appreciated for both their fragrant flowers and their fruit. One recent suggestion is that there are three basic citrus species, from which sprang the wide variety we enjoy: the citron (*C. medica*); the mandarin (*C. reticulata*); and the pummelo (*C. maxima*).

Citrus moved west after Alexander the Great discovered them in India in the 4th century BC, although the citron was known in Southwest Asia before 400 BC. This resembles a large lemon and became important in the Jewish Feast of the Tabernacles; Etrog, a small-fruited variety, is grown especially for the Feast. Although now only a minor member, citron has given its name to the whole genus, indicative of its historical importance.

Other citrus fruits, probably limes and lemons, arrived in Europe in the early Christian era, although they were still exotic and available only to the wealthy. The Arabs prized them and introduced bitter oranges, lemons and limes into areas they conquered, including Spain. Then, as now, citrus trees needed two things: plenty of water and no serious frost. This means that they require irrigation if grown in dry areas, and protection in climates subject to occasional frosts. As rich northern Europeans became fascinated with their taste and the delicious scent of their flowers, they constructed orangeries and large hothouses in order to grow them. The elegant Orangery at Kew Gardens, southwest London, was built when

Opposite Some of the varieties or species (he was undecided) of 'sour lemons or limes' of India drawn at the request of William Roxburgh, perhaps during his superintendency of the Botanic Garden in Calcutta (1793–1813). During his tenure, his Indian artists produced some 2,500 botanical illustrations. One set remained in Calcutta while the second came home to the Royal Botanical Gardens, Kew.

Tangierine Orange

Above left 'Limon cedrato' or citron. Johann Christoph Volkamer, a wealthy merchant at Nürnberg (Nuremberg), established orchards and an orangery for his extensive collection of citrus trees. These were recorded in his 2-volume *Nürnbergische Hesperides* (1708–14), with over 100 plates, in which botanical art met Rococo garden style.

Above right The tangerine orange (*Citrus tangerina*): an easily peeled variety associated with the region around Tangiers in North Africa. It was commended in the 19th century by Mr Tillery, gardener to the 5th Duke of Portland, as the tastiest orange he had known grown in Nottinghamshire, in England. The Duke's Welbeck estate had brazier-heated walls in its extensive kitchen garden and glass houses.

the gardens were still the private preserve of King George III. Unfortunately, the light levels were insufficient for successful citrus growing.

Although citrus fruits proved difficult to grow at home, British, Dutch and other northern European seafarers encountered them in their travels, and were grateful for them. Long sea voyages, accompanied by monotonous and nutritionally inadequate victuals, meant that scurvy was a major hindrance to European imperial and commercial aspirations. Scurvy causes bleeding from the gums and under the skin, weakness, diarrhoea and emaciation; it can kill. We don't know who first recognized that citrus fruits could actually cure this dreadful disease, and that regular consumption could prevent it occurring in the first place. However, by the late 15th century, the value of eating citrus fruits in the cure and prevention of scurvy was mentioned, and some ships' captains were aware of this before it became routine medical practice. In the mid-18th century the Naval Surgeon James Lind famously tested citrus along with other recommended remedies for scurvy in an early clinical trial. He found that lemon was especially effective, although it took another three decades before the British Navy made provision for the supply of limes or lemons to sailors.

Citrus fruits contain variable amounts of ascorbic acid, the vitamin that prevents (and cures) scurvy, which humans cannot make in their bodies. Ironically, the lime, which gave British sailors their nickname 'limeys', is a less reliable source than some other citruses, though many other fresh fruits and vegetables also contain it. The variable amount of ascorbic acid in foodstuffs complicated the evaluation of citrus fruits in the scurvy scenario, but the health-giving properties of the fruits are still prominent in modern marketing. From the 1980s, the Nobel Prize-winning chemist Linus Pauling advocated high doses of ascorbic acid for the treatment of the common cold as well as the prevention of cancer and cardiovascular diseases.

Sweet oranges (*C. sinensis*) were cultivated in Europe by the 15th century, and Columbus took them and lemons (*C. limon*) to the Americas on his second voyage. There they were grown on several Caribbean islands, as well as in Florida, which is still one of the centres of world production. California became Florida's major North American rival, after groves were established there in the 19th century. Just a year after the transcontinental railway reached Los Angeles, a citrus grower successfully shipped a load of oranges back east. Other major producers include Brazil, several southern European countries and Israel. Sweet oranges are now grown principally for their juice, which modern methods of preparation and transport have made readily available.

Continuing crosses have increased the variety of citrus fruits on the market, which can now be purchased all year round. At the same time, intensive cultivation has inevitably brought with it the perennial problems of pests and diseases. Freak frosts in generally warm climates also affect the annual harvest, and protection via sludge pots (now often banned because of pollution) and spraying trees with water when frost threatens are the two major strategies of growers. Spraying relies on the fact that as long as water (rather than ice) is present on the leaves, the leaf temperature will not fall below the freezing point of water.

'Hesperidium' is the scientific term for the whole fruit, after the golden apples of Hesperides sought by Hercules, which were neither orange nor apple. If the juice and flesh (pulp) are what we most readily consume, the peel is just as valuable. The peels of various citrus fruits contain many oils that are widely used in soaps, perfumes and cooking. The bergamot orange is grown exclusively for the oils in its peel which give Earl Grey tea its distinctive flavour, and the peel of Seville oranges makes the best marmalade.

'De Aloe' from Leonhart Fuchs'
De Historia Stirpium (1551), one
of the great herbals of the 16th
century, illustrated with fine-
quality woodcuts. Fuchs may
have grown his aloes in pots at
the botanical garden he created
as part of the university at
Tübingen. In the 1st century AD,
and further south, Pliny reported
the use of special conical pots
for aloes.

Aloe

Aloe perryi, A. vera, A. ferox

The Succulent and Its Healing Gel

*You ask me what were the secret forces, which sustained me during my long fasts.
Well, it was my unshakable faith in God, my simple frugal life style, and the Aloe
whose benefits I discovered upon my arrival in South Africa.*

Mahatma Gandhi (letter to his biographer Romain Rolland)

Medicinal aloes were so prized that Aristotle advised his one-time pupil Alexander
the Great to take control of the island of Socotra in the Indian Ocean, where the
best aloes (*Aloe perryi*) grew. Or at least that's how the story goes. Perhaps the
human worth of a plant might be measured by the quality of the legend it attracts.

Aloes are succulents, plants that have various adaptations to cope with aridity.
The aloes' modified leaves act as water storage organs during the dry season, and
the liquid they contain, as well as the tissue, attracts herbivores – in southern
Africa elephants have a penchant for them. Probably in order to deter such interest
the leaves have protective prickles, and many species produce bitter-tasting chemicals including the aloins. The amount of aloin (and other bioactive products)
varies from species to species, plant to plant, with age and season, and in response
to damage. This can lead to inconsistencies in the quality of aloe products and is
one reason why it has proved so difficult to assess their reputed medicinal benefits
and gauge toxicity.

The name aloe is derived from the Arabic *alloeh*. Rather confusingly the same
word was used for the commodity the plant yielded, aloes, in the form of shiny
blackish bricks or lozenges. When a leaf is cut, latex is exuded from special cells just
under the skin; this was heated and condensed until it solidified. The easily transportable lozenges could then be melted again as needed and were often combined
with other drugs. The Egyptians used aloes as early as the mid-2nd millennium
BC, though Hippocrates did not mention the drug. It entered the Roman pharmacopeia only in the 1st century AD – Pliny reported that the leaves of the Asiatic
kind were applied fresh to wounds. This use is still promoted, although clinical
evidence is equivocal. The best aloes, he said, came from India, for which we can
read the Indian Ocean trade routes. He extolled it principally as a laxative, though
listed many other situations in which it was effective, including against hair loss.

Wound healing and purging were the main uses of aloes, although it was recommended for numerous therapeutic, cleansing and moisturising tasks, internally
and externally. Its popularity was especially strong in the Arabic medicine of the
eastern Mediterranean, exported by the Muslim conquests. Over time the plants,

as well as the condensed drug, were exported to the east as far as China and into Europe. Everywhere it gained popularity and filtered into the pharmacopoeias.

Africa, especially the south, houses *Aloe* hotspots, with a fabulous diversity of species. At the continent's southernmost tip, *A. ferox* was a long-standing indigenous favourite. Today it is the source of an important local industry as the demand for aloe products continues to expand; traditional ways of managing and harvesting leave the plant intact for further cropping. Africa's output of *A. ferox*, however, is dwarfed by that of *Aloe vera* from Mexico, the southern US and parts of South America. *A. vera* came from the Arabian Peninsula, although quite where is unclear. It was taken to the New World by Columbus, and spread from the West Indies. Aloe latex and its products are not benign cure-alls, but recent studies do point towards a role in lowering blood sugar in diabetics and blood lipid levels in patients with an excess.

Aloe ferox, the 'great hedgehog aloe', from *Curtis's Botanical Magazine* (1818). This South African aloe grows to 2 m (over 6 ft) in height, with leaves reaching 1 m (3 ft) in length. Traditionally a few leaves are taken from multiple plants and arranged around a lined depression to collect the brown exudate dripping from the cut surface. Concentrated by heating and then dried, the resulting solid bitter aloes have been used locally for centuries.

Mexican Yam

Dioscorea mexicana, D. composita

Making 'the Pill'

No woman can call herself free until she can choose consciously whether she will or will not be a mother.

Margaret Sanger, 1919

There are many pills, but only one is '*the* pill'. The oral contraceptive or birth control pill was known simply as this soon after its introduction in 1960. Widespread availability for unmarried women came too late for the freedoms of the 'swinging sixties', but it would in time offer unprecedented childbearing choices for sexually active women. The pill is unusual in that it is not taken to cure or alleviate sickness, but to prevent a natural process. The body's reproductive hormones control a woman's menstrual cycle through a delicate series of feedback loops. The pill's daily dose of the hormone progesterone interferes with this and suppresses ovulation. The original source of the hormones used were plants – two non-edible species of Mexican yams known locally as *cabeza de negro* (*Dioscorea mexicana*) and *barbasco* (*D. composita*).

Research in the 1930s and 1940s involving steroids and sex and other hormones was exciting and full of medical promise. While biologists determined how these chemical messengers controlled many of the body's functions, chemists tried to find ways of supplying large amounts at minimal cost, for research purposes but especially for therapeutics. Plants have steroids too and maverick organic chemist Russell Marker was the first to realize the potential of the Mexican yams. His genius lay in an uncanny ability to synthesize analogues of the complex organic compounds found in nature, provided he could find the right raw material. Impressed with Japanese work with the steroid diosgenin from yams, he put his energy into finding a cheap source of plant steroids.

With the help of botanists and knowledgeable locals, Marker's team analysed some 400 plants collected during trips in the early 1940s to the southern US and Mexico. Marker became interested in the *cabeza de negro* yam. This large, coarse vine with heart-shaped leaves and huge tubers grew wild in the stifling forests of Veracruz, eastern Mexico. Marker persuaded a local shopkeeper there to find some tubers for him, and from this diosgenin-rich source he produced enough progesterone to warrant industrial-scale exploitation. As human steroids all share a similar basic structure, synthesis of one opened the way to the manufacture of other sex and adrenal gland hormones.

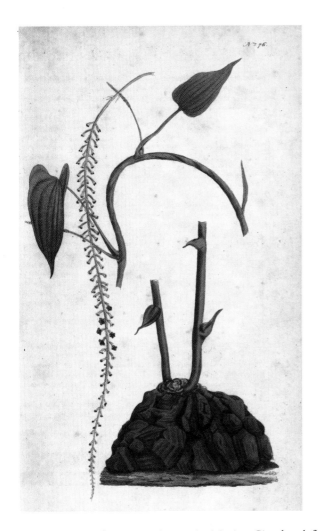

In 1944 Marker set up Syntex in Mexico City, but left early in 1945. He had determined that the *barbasco* yam had a higher yield of diosgenin and reached productive size in about three years (*cabeza de negro* required six to nine years). Using this plant, Syntex chemists Carl Djerassi and Louis Miramontes improved the processing of the tuber and the means of producing large quantities of pure progesterone. They went on to produce an oral, rather than injectable, form of progesterone – norethisterone. Tested for its efficacy to treat menstrual disorders and miscarriage, its potential as an oral contraceptive became apparent. Clinical trials in Puerto Rico followed and Enovid was the first pill marketed.

Syntex, the pill and supplies of diosgenin established Mexico as the world's leading producer of high-quality plant-derived hormones. There were many positives, but also negatives. Unemployed peasants collected the yams in the jungle, enduring hard conditions before selling them on to middlemen. Such collecting was unsustainable. Price increases in the 1970s, in part a result of the Mexican government nationalizing yam gathering, made other options such as total synthesis financially viable.

Dioscorea mexicana: its characteristic above-ground tuber is found in other yam species, notably the 'elephant's foot' yam (*D. elephantipes*) of South Africa. Rich in saponins, this is also threatened by over-collection for garden specimens and by local people who use it medicinally.

Madagascar Periwinkle

Catharanthus roseus

Delicate Flower, Powerful Treatment

This plant deserves a place in the stove [hothouse], as much as any of the exotics.
Philip Miller, 1768

Madagascar is a special island, home to a diverse, distinct flora and fauna. Some 80 per cent of its native plants lived nowhere else before being transported to other lands. This includes the Madagascar or rosy periwinkle. Popular as a tender bedding annual, it is pretty without being sumptuous, more bridesmaid than bride. Yet within its tissues the periwinkle contains the unique alkaloids vinblastine and vincristine that have changed the face of childhood cancers.

On its island home and elsewhere the periwinkle had also gained a reputation as a medicinal herb. In Jamaica 'periwinkle tea' was used as a remedy for diabetes. Insulin had transformed this disease, but was expensive and had to be injected. The efficacy of the periwinkle leaves was tested using a sample received from Jamaica in 1952 in the laboratory of Robert Noble at the University of Western Ontario in Canada. Effects on blood sugar and glycogen levels using oral preparations were disappointing. Fortuitously, along with blood tests for glucose Halina Czajkowski Robinson ran white blood cell counts on her own initiative. The results of her tests revealed that something in the periwinkle was suppressing the production in the bone marrow of these vital cells of the body's defence system. In 1954 biochemist Charles T. Beer joined the team and four years later succeeded (after some 5,000 attempts) in isolating the 'something'. It was an alkaloid named vinblastine, which could be tested in tumour cell cultures on the laboratory bench. Although Beer's method of producing vinblastine would be patented, parallel work was underway at the giant pharmaceutical firm of Eli Lilly in Indianapolis, USA. Here they were conducting a sweep of analyses of plants for potential drugs and had heard about periwinkle and diabetes, this time from the Philippines.

Late in 1959 vinblastine moved from the culture dish to patients. It was made commercially available in 1961 (as Velbe) and joined in 1962 by the similar but more active alkaloid vincristine (Oncovin). In combination therapy, despite side effects, this became the bastion of the treatment of childhood leukaemias. Its use has been extended to some lymphomas and lung cancers, Kaposi's sarcoma in HIV/AIDS and autoimmune conditions. The periwinkle drugs work as 'spindle poisons', interfering with the formation of the microtubules of the spindle, the

apparatus in the dividing cell that helps move the chromosomes around during cell division.

Symbol of hope for these diseases, the periwinkle has also become something of an emblem for biopiracy too. Many deplore the exploitation of the knowledge and resources of the developing world by the developed. Madagascar, the evolutionary home of the periwinkle, seems to produce the best quality alkaloids, but the samples of periwinkle plants in this story were from elsewhere. Various countries today provide the raw material, which is needed in large amounts, but the rural Malagasy people who collect the wild plants or grow them on a small scale continue to receive poor prices. There is also a fear that global habitat destruction reduces the chances of finding other plants akin to this life-saving flower.

The Madagascar periwinkle was first grown as an ornamental in France, at the royal gardens of Versailles and Trianon. Seeds from there were sent to Philip Miller at the Chelsea Physic Garden in the 18th century, who extolled their summer-long profusion of blossoms. Apart from its need for winter warmth, its cultivation requirements are easy to satisfy, and the periwinkle naturalized wherever it was taken in the subtropics and tropics.

Technology and Power
The Material World

Above Quercus squamata is now known as *Lithocarpus elegans* and is no longer considered to be an oak. In the 19th century it was described as a large timber tree from the species-rich subtropical Garrow (Garo) mountains in northeast India; the wood was said to be lighter coloured than English oak, but equally strong and close-grained. The name may have changed but the functionality remains.

Whoever controlled the lands where the Cedar of Lebanon grew controlled access to this most valuable of trees and the bounty that its wood afforded. Much in demand as a durable and fragrant timber for Phoenician ships as well for construction, it provided a valuable commodity for exchange. Its resin, too, was held in high regard. Cedar wood thus sets the tone for Technology and Power, examining the material applications of plants and what benefits that might bring.

What cedar had done in the eastern Mediterranean, oak would do for various seafarers in Europe. Power at sea was matched by majesty at home. Oak provided the durability and strength to create many great churches and other buildings. Yew conferred power differently. It made stout hunting spears that have lasted long enough to date their use back some 450,000 years. The subtle difference in flexibility between sapwood and heartwood created the longbows that helped the English triumph at Agincourt in the 15th century. And yew's biochemistry has yielded weapons of another sort – drugs to kill the rapidly dividing cells of certain cancers.

All these woods can also be used for furniture too, but there is always the shock of the new. When Europeans encountered the tropics they found huge trees, their growth stimulated by the moist, hot conditions, and untouched by loggers. These provided usable wood in dimensions hitherto unknown. Tropical hardwoods, mahogany in particular, became the wood of choice for furniture

in the 18th century. Easy to work and inherently stable, its products were highly desirable. By the time the fashion for mahogany had run its course, the future of the tropical woodlands was already threatened.

From tree to grass – the amazingly versatile bamboo. Spread through the temperate and tropical world, these perennial, woody, hollow-stemmed plants have proved to be extraordinarily adaptable in human hands. Their stringy strength lends them to house building in Japan. Lashed together lengths of the giant bamboos created buoyant rafts. Twisted bamboo reminiscent of a steel cable acted like one in Chinese suspension bridges. From baskets to fabric, bamboo fibres have been woven into objects and clothes as this plant continues to meet our needs.

Today bamboo can be mixed with other, older fibres such as cotton, hemp and linen made from flax. Wool from animals is also used of course, but vegetable fibres offered something different. From these came string, cord and cloth, clothes and sails, paper and furnishings, the industries of spinning and weaving, high fashion and the hangman's rope. All these plants contribute the raw materials of our material world.

Above left The majestic Cedar of Lebanon (*Cedrus libani*), with details of the needles and cones – male above, female below – and winged seeds, which are released from the mature cones.

Above right Carding the cotton, an essential stage in its preparation for spinning, as recorded by Pierre Sonnerat in his *Voyage aux Indes orientales et à la Chine* (1782).

Cedar of Lebanon
Cedrus libani

Foundation of the Phoenician Empire

My friend up to now the high-grown cedar's tip would have penetrated to heaven.
Make from it a door whose height will be six-dozen yards.

Epic of Gilgamesh, Tablet V

Towards the higher reaches of the steep rocky mountainsides of Lebanon and Syria are small stands of majestic conifers, the Cedars of Lebanon, which can grow to some 40 m (130 ft) in height. They are the descendants of trees which, along with cypresses, junipers and other pines, once clothed the slopes of the Lebanon and Taurus mountains in evergreen beauty. If modern Turkey is the best remaining source, it is Lebanon that has claimed this iconic tree as its national symbol.

Measured on a geological timescale, the *Cedrus* genus is a relative youngster. It evolved in the early Tertiary period (65 to 55 million years ago) and fossilized remains are hard to distinguish from the living species. In terms of human history it was a crucial tree for the ancient civilizations that occupied the Levantine coast. Cedar wood provided the late Bronze and Iron Age Phoenicians with a most valuable commodity, which their neighbours – the Egyptians, Assyrians, Israelites, Babylonians, Persians – would seek to attain by trade, demand as tribute or take by force. The construction of a coastal railway during the First World War and its subsequent fuelling reduced much of what then remained of Lebanon's significant stands of cedar to ash. This was a sad finale for trees once so prized that their wood was used for the coffins of Egyptian pharaohs.

Cedar's essential oils imbue it with a highly desirable scent. It is also an enduring one, as archaeological finds of still fragrant wood attest. Other chemicals in the resin impart good protection against wood-boring insects and microbial decay. Unlike the spreading contours of landscaping cedars planted in European parks from the 18th century onwards, those growing more densely on home soil were taller and straighter. They yielded impressively long runs of usable wood much prized in the ancient world for the construction of major buildings and ships.

Solomon famously sought cedar for his temple and royal palace in Jerusalem. He negotiated with Hiram, king of Tyre, for timber and skilled woodworkers, offering in return silver and vast amounts of olive oil and wheat. When the temple of Amun-Ra wanted cedar for a new sacred barge for the god in the 11th century BC, Wen-Amon, a senior temple official, sailed from Thebes to the Phoenician city of Byblos. Egyptian gold, silver, linen and 500 rolls of papyrus were the price.

The Phoenicians were famed merchants and shipbuilders. Their craft, made of timber from their great cedar groves and equipped with a single sail and oars, sailed throughout the Mediterranean using direct open-sea routes. As well as building their ships from cedar trees they also traded in the highly prized cedar logs.

Cedern-Gruppe am Libanon.

The Phoenicians were master shipbuilders. Their slow but capacious merchantmen sailed the breadth of the Mediterranean. Famed for their length, cedar wood planks (and masts) offered strength and flexibility too. Specially thickened cell walls are laid down on the underside of the trunk, which in life help to keep the trunk vertical against the heavy pull of the branches. Such dense wood absorbs less water when immersed, aiding the water-resisting qualities of the resin. In such ships the Phoenicians exported their luxury goods – carved ivories, jewelry made from precious metals – and traded in raw materials such as copper ingots, towing behind them the prized cedar logs from home port to distant destination.

An ancient cedar grove in Lebanon. The Forest of the Cedars of God (Horsh Arz el-Rab) in northern Lebanon is part of a UNESCO World Heritage Site that celebrates and conserves the ancient lineage of *Cedrus libani*. The trees may be relics of populations of the *Cedrus* genus that ran along the mountain ranges here before the Tethys Sea closed and became the Mediterranean.

Oak

Quercus spp.

Might and Majesty

Leaves and acorns of the European white oak, *Quercus pedunculata*, from *The North American Sylva* (1865). It can take 50 years before a tree produces acorns; then, with the assistance perhaps of a helpful squirrel or jay, it takes two years for a small shoot to develop from a buried acorn, since the root system is the sapling's priority.

Great oaks from little acorns grow.
Late 14th-century English proverb

Quercus (oak) is a large genus of up to 500 species, distributed mainly over the northern hemisphere, but with a native species in the Colombian Andes. The trees can live from sea level to as high as 4,000 m (over 13,000 ft) and come in a variety of guises including some evergreen and semi-green species, but it is the tall, stately deciduous trees that have been most prized. Slow-growing hard woods, they generally have thick, gnarled bark, rich in tannins. These naturally occurring chemicals exploited for preparing or tanning leather and maturing wine also make the wood especially good as a building material, as they deter insects and resist rot. The thick-walled fibrous cells of the oak's core give it strength.

Oaks probably originated in Asia and were widespread 55 million years ago; their modern distribution reflects the ecological and climatic changes that have occurred during recent geological periods. They are wind-pollinated, and germination, as with many plants, is a hit and miss affair. Squirrels often help to distribute the acorns, hiding them in the ground and then forgetting where they left them (or meeting with an accident). Most oak species need temperate weather and relatively rich, moist soil. When conditions are favourable, individuals can live for several hundred years and reach enormous sizes.

Different species have been exploited for a variety of uses. A favourite was *Q. robur*, native to Britain and much of northern Europe. The material of choice for ships until the coming of metal hulls, it helped create British supremacy at sea. Several characteristics were needed for wood for shipbuilding: large trunks from which to hew timber, and malleable planks that were light and watertight. Oak had all these. Although it was dominant in British forests, such was the demand for ships, houses, furniture and firewood that much had to be imported from Scandinavia and other Baltic countries from the 17th century. The most common American species, *Q. alba*, the white oak, helped make the fortune of the Massachusetts Bay Company, which shipped timber home in the ships that brought settlers across the Atlantic. Another American species, *Q. virginiana*, the 'live oak', was considered the most durable of all, although it is less commonly used now.

Yet another species of oak has a special place in the affections of wine lovers: *Q. suber*, the source of cork. The bark of this semi-evergreen tree is especially thick and the ancient Romans used it in a variety of ways: for insulation, shoe soles, anchor floats and to stopper bottles – all still familiar today. The bark itself is probably an evolutionary adaptation against fire; it can be stripped every ten years, after the bark has regenerated, and the tree can live for 100 years or more.

Although acorns are edible (after leaching out the bitter tannins), and were a staple of many Native Americans until the late 19th century, especially in California, they are not much consumed today. However, they still form the basis of food for pigs in Italy and Iberia, from which famous hams are made.

Two East Asian oak species, with their leaves, acorns, bark and wood: *Quercus aliena* (right), oriental white oak, a useful timber tree; and *Quercus variabilis*, Chinese cork oak (left), which has lower yields but similar characteristics to European cork oak, with cells coated in a waterproofing wax.

Yew

Taxus baccata, T. brevifolia

Medieval Longbows, Modern Medicine

A sprig of *Taxus brevifolia*, which grows in the mountains of the western coast of North America from southeast Alaska to northern California. Native Americans used the wood for canoe paddles, bows and spears. Until the development of taxol, the forestry industry regarded yew as a nuisance, cutting and burning these ancient trees of the understorey.

We few, we happy few, we band of brothers.
William Shakespeare, *Henry V,* Act 4, Scene 3

On the morning of St Crispin's Day, 25 October 1415, an English army led by Henry V faced that of France under Charles VI. By the end of the day Henry was victorious and the legend of Agincourt had begun. The English were not as out-numbered as the old chronicles would have it, but had far more archers equipped with their longbows made of English or European yew (*Taxus baccata*). It was these devastating weapons, strategically deployed, that made the difference.

Yew is an ancient material of aggression. Hunting spears are among some of the most ancient wooden artifacts discovered. A yew spear tip from Clacton-on-Sea, Essex, England, dates from around 450,000 years ago. A spear lodged between the ribs of a mammoth in Lower Saxony, Germany, is some 90,000 years old. Ötzi the Iceman carried an unfinished yew bow on his final journey in the Ötztal Alps, on the border of Italy and Austria, over 5,000 years ago.

By the reign of Henry VIII of England the longbow's heyday was waning, but it still had a presence. Recovery of Henry's warship the *Mary Rose,* which sank in 1545, has yielded a considerable shipment of yew bows and staves. The quality of the wood indicates that it was imported European or perhaps western Asian timber. Venetian merchants who controlled so much of the trade at the time sourced the wood. A specified number of staves were set as a tax on each barrel of wine imported into England.

Other woods were also used for bows, but yew is special. If the bow staves are cut radially from a length of even, straight yew timber, the result has two natural layers. The pale sapwood from just under the bark forms the flat outer layer, away from the archer; it resists being put under tension. The rounded belly of the bow is the heartwood layer, which resists compression. This resistance is stored as tre-mendous energy as the archer draws the bow – energy that propels the arrow upon release. Strong and amazingly elastic, the water-conducting cells of yew wood – tracheids – have a helical thickening that acts like a series of coiled springs.

Yews are long-lived trees with remarkable powers of regeneration, but the need for bow staves took its toll. So too did the clearance of forest for agricultural land. Some four hundred years after the *Mary Rose,* and across the Atlantic Ocean, the

Pacific yew would face similar pressures, but this time it involved metaphorical warfare between cells of the body.

In 1962 bark was stripped from mature yews of *Taxus brevifolia* in Washington State and screened as part of a large project to harness the potency of plant products as cancer drugs. Once the active principle had been isolated and its structure and mode of action were determined, clinical trials were run. Thirty years after the bark had been harvested, in 1992, taxol (trademarked as Taxol® in 1994) was licensed for advanced, refractory ovarian cancer. Today variations on the original drugs are also used to treat forms of breast, lung, head and neck cancers, and the AIDS-related Kaposi's sarcoma.

The taxol is found in the bark of mature Pacific yews, a large amount of which is needed to produce a small quantity of the refined drug; obtaining the bark meant killing the tree. In 1990 after the realization that clinical success would lead to huge demand, environmentalists joined with the American Cancer Society and two key scientists of the taxol story to call for sustainable exploitation of the yew by having it listed as a threatened species. The petition failed, but the Pacific Yew Act went some way to protecting the trees. Since then part-synthesis of taxol has made the raw material go further. Means of producing drugs from clippings, twigs and needles, including from European yew, and the use of plant cell cultures have also been developed. Sadly, exploitation of the Himalayan yew (*Taxus contorta*) for taxol in northwest India and western Nepal has led to a 90 per cent decline in the number of trees. It is now listed as an endangered species on the IUCN's Red List.

The winged god of love Armour tutors the Lover in this illustration from the medieval allegory of chivalric love, the 'Roman de la Rose'. Armour is equipped with a longbow, which is drawn ready for release. Huge forces were required to draw the bow – bodies of archers found on the *Mary Rose* have characteristic asymmetries in the bones of the left (bow arm) and right (drawing arm) shoulder and elbow.

Flax

Linum usitatissimum

Linen and Lino

Love is like linen often changed, the sweeter.

Phineas Fletcher, *Sicelides* (performed 1614), Act 3, Scene 5

Flax has clothed humanity since at least the 4th millennium BC, though it was brought into the tilled field initially for its oil-rich, edible seeds. Probably domesticated from pale flax (*L. bienne*) about 8,000 years ago, its consumption is still older.

A key event in the domestication process appears to have been a genetic one that increased the unsaturated fatty acid content. Exposed to air, linseed oil slowly oxidizes and solidifies. Applied to densely woven linen, it strengthened ancient body armour. The same principle allowed pigments mixed with linseed oil to dry in layers when painted on to linen canvases. And what served high art also protected woodwork against the elements. In the 1860s Frederick Walton developed linoleum or 'lino' flooring by building up layers of oxidized linseed oil mixed with cork or wood dust on a fabric backing. After the discovery of germs in the 1880s, housewives were encouraged to throw out their carpets and lay linoleum. It was easy to clean and the linseed oil was also touted as an antibacterial agent. Back in fashion after being overtaken by vinyl in the 1950s, lino is now a 'green' product.

Compared to the oil plants, fibre-producing flaxes are taller (around 1.2 m/4 ft) and leaf only at the top. Linen was both a utility and a sacred fabric in ancient Egypt. Easily bleached and robust enough to stand frequent washing, pure white linen was worn by temple priests. Egyptian tomb paintings of the early 2nd millennium BC show flax being turned into linen. Processing the plants involves first retting (rotting) in water and then scutching (beating), releasing the bast fibres (just beneath the outer layer) for spinning and weaving. Flax fibres were made into string, rope and sailcloth (flax is about twice as strong when wet) as well as cloth to be worn.

In cooler northern and western Europe where flax grew well, linen made a useful layer beneath wool, being smooth on the skin and absorbent. Cotton and silk were pleasing, but prohibitively expensive outside their native regions. Linen was woven in the home for domestic use, and in the urban prosperity of the late Middle Ages high-end linens in an expanding range of styles and weaves were traded. From bed-sheets to altar-cloths, fine linen found a place.

More linen on the market meant more waste cloth at the end of its use. The technique of making paper from rags completed its slow journey from China to Europe via Muslim Spain in the 12th century. From the 16th century onwards, various centres of linen weaving – Bruges, Antwerp, Belfast – developed distinctive fabrics and reputations, while Russia grew much of the raw material. Trade and war at sea increased the need for both sailcloth and sailors' slops. Linen found new uses as warfare changed. In the First World War, linen held the ammunition on machine gun belts; waterproofed it covered aircraft frames.

Linen's failing perhaps is its inelasticity which makes it crease easily. The advent first of cheap cottons and then synthetic fabrics caused a precipitous decline in linen's place in the clothing market. New blends, expensive price tags and rarity value have created a new aura around this ancient cloth. And flax foods are in fashion again, marketed for their alpha linolenic acids, which the body converts to omega 3 fatty acids thought to prevent cardiovascular disease.

Hans Christian Andersen's fable of metamorphosis, *The Flax* (1843), takes this most useful of plants on a journey from field to fabric, from ragbag to paper and a final immolation in the fireplace. Cheerful stoicism might be Andersen's intended message, but he neatly captures *Linum usitatissimum*'s utility.

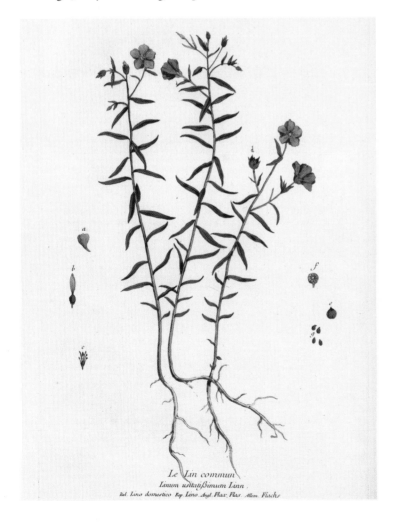

Le Lin commun.
Linum usitatissimum Linn.
Ital. Lino domestico Esp. Lino Angl. Flax. Flas. Allem. Flachs

Hemp

Cannabis sativa

Textiles and Old Rope

… three merry boys are we,
As ever did sing in a hempen string
Under the Gallows-Tree.

John Fletcher (with Ben Jonson & others), *The Bloody Brother* (c. 1616)

Planted as an annual, cultivated hemp plants can reach 5 m (over 16 ft) in height. From these tall stems come the bast fibres used for making textiles and paper, as well as inestimably useful string, cordage and rope, while the seeds yield an edible oil. Fibres from different plants have been referred to as 'hemps', but the real thing comes only from *Cannabis sativa*. Shorter and leafier varieties of the same plant are the source of psychoactive cannabinoids. These are concentrated in the female flowers, which, along with nearby leaves, are processed as hashish and marijuana.

Both the stem fibres and the cannabinoids have been used for a very long time in the plant's homelands of Central and northeastern Asia. Chinese shamans employed the narcotic properties for medical purposes and the mind-altering effects. Knowledge of these uses may have spread westwards from China into India or may have arisen independently. Neolithic pottery from China, the 'land of mulberry and hemp', may well bear the imprint of hemp cloth. Mulberry – the food of silkworms – clothed the rich, while the poor made do with hemp. Under Confucian influence filial duty required surviving children to mourn their parents in hemp clothing, regardless of wealth. Han dynasty China (206 BC–AD 200) witnessed the paper-making revolution, using new and recycled hemp fibres (clothing and fishing nets) and mulberry bark.

After the plant's introduction from Asia, hemp rope and cordage appear to have been taken up quickly in the Mediterranean during the Classical period before spreading inexorably westwards. Trade, exploration and conquest all relied increasingly on hemp as ships increased in size and complexity. Hemp (and linen) sails and rigging help propel the ship; sailors slept in hammocks of hemp canvas. Tremendous hawsers towed ships and hemp ropes anchored them, while a hemp-line gauged the water's depth.

Early settlers in Virginia in the New World were obliged to grow hemp, the land and climate being suitable. Much later, American 'duck canvas' shaded the covered wagons that headed west to open up the country in the 19th century. But it was Russia that met the growing naval demands. Peter the Great saw the opportunity to

Tab. 706

The hemp plant, *Cannabis sativa*, is the source of many useful products as well as a narcotic. Ropes made from the strong stem fibres were crucial to sailing ships before synthetic versions and were tarred to prevent deterioration. When they did wear out, they were unpicked and the resulting oakum was used to maintain the caulking on hulls and deck planks. The unpleasant task of picking oakum was assigned as a naval punishment.

CANNABIS SATIVA.
Der zahme Hanf.

turn some of Russia's vast land and its serf population to the cultivation of hemp. This was hugely successful, and hemp became a major export crop.

Hemp was crucial until the end of sailing ships and the rise of synthetic fibres. Its modern history has been dominated by its use as a narcotic, which has seen its growth and consumption outlawed. Varieties with low THC (tetrahydrocannabinol) are now being grown commercially, as hemp oil and seed make a health-driven comeback and the fibre's green credentials are increasingly appreciated.

= G. hirsutum Lin. Sp. Pl.
forma religiosa Roxb.
fs.

No. 1497 Gossypium religiosum Willd

1497 Gossypium fuscum R.

herbaceum   G. religiosum, Roxb.
= G. barbadense var. religiosum, Mast.

Cotton

Gossypium spp.

Clothing the World

Cotton is King
David Christy, 1855

The ubiquitous popularity of jeans ensures that worldwide demand for cotton remains high. It comes at a price, however, as modern methods of growing use large amounts of fertilizers and pesticides, mostly petroleum-derived, as well as a lot of water. In warm climates cotton is a perennial and can grow into a tree, but it is generally treated as an annual, when it is a large shrub of up to 2 m (over 6 ft) high. It needs rich soil, with plenty of rain in its growing season and a dry spell for harvesting. Of the around 50 species of the genus *Gossypium*, only four have been exploited for the plant's fibres, which are found in the ripe seed pod or boll. Two New World species, *G. hirsutum* and *G. barbadense*, were used in pre-Columbian times; two Old World ones, *G. arboreum* and *G. herbaceum*, were also long prized.

The origin of the two Old World species is obscure, as their wild ancestors are not yet identified, but they probably came from Africa. Archaeological evidence for cotton in the Indian subcontinent dates from about 2500 BC. The Greek historian Herodotus described both the plant and Indian weaving of the strands; Indian writings mention it even earlier. The word 'cotton' is of Sanskrit origin. The Arabs introduced cotton to Sicily and Spain as they expanded their empire from the 8th century. However, European weaving and spinning techniques did not match those achieved in India, where muslins and chintzes of great delicacy and beauty were produced. Most early European cottons were actually fustians, a mixture of cotton and flax.

New World cotton had already established itself in pre-Columbian societies. *G. barbadense* was native to Peru, where it was an important trade commodity between coastal communities, which needed it for fishing nets, and upland ones, where it was grown. In the 1st millennium BC the Paracas culture in Peru created elaborate textiles from camelid wool woven with cotton. 'Egyptian cotton', although now widely grown there, is actually the New World species *barbadense*. *G. hirsutum* was probably domesticated several times, but especially in Mesoamerica. The Spanish invaders in the early 16th century abandoned their scratchy woollens and linens for the softer cottons that the Aztecs wore. This species now accounts for almost 90 per cent of world cotton production.

Opposite The flowers, ripening boll and seeds of *Gossypium religiosum*. As the handwritten notes testify, this South American cotton has been renamed several times. William Roxburgh commissioned this drawing in India, where the plant had gained a reputation for being cultivated by mendicants or found near temples.

It took a while for cotton to become common in Europe, but when it did it transformed the textile industry. And since cotton is much easier to wash than wool or linen, it also had important consequences for cleanliness and public health. Originally most European cotton was imported in its raw state from India, where the East India Company enjoyed a monopoly. In the 18th century, innovations in harnessing power and machines to separate boll and seed and for combing and weaving the fibres revolutionized cloth production in Lancashire in the north of England.

American ingenuity also improved cotton processing. Eli Whitney's cotton gin of 1793 allowed the much quicker mechanical separation of the fibres from the seeds in the boll. These and other developments brought the price of cotton fabrics down, which in turn stimulated demand. The United States began to rival India in supplying the commodity. Climatic conditions were favourable in the southern states, where large-scale cotton growing began in Tennessee in 1807 and soon spread. African slaves supplied the labour in the plantations, continuing the inhumane strategy that had given sugar cane its place on world markets. American exports to Britain increased hugely by 1853, much to the detriment of India.

The American Civil War (1861–65) disrupted the supply of cotton, a gap that India only partially filled. It led to unemployment in the Lancashire mills and divided British sympathies about the War, despite the fact that slavery had been one cause behind the conflict. The defeat of the South ended slavery and punctured the old plantation system, but didn't stop cotton monoculture there, and with it the seemingly inevitable problems of pests and diseases. From the late 19th century the worst of these was the boll weevil, the female of which lays its eggs on the cotton plant's buds.

The ongoing struggle between insect and agribusiness has involved pesticides and genetically modified (GM) plants. Between these and chemical fertilizers, as well as heavy use of water, cotton has come in for much criticism from environmentalists. Demand remains high, however. Cotton seed and its oil have myriad uses, from animal food to paints. Recycled fibres are used in the paper for American currency. Short fibres go into making cellulose used in explosives, shoes, handbags and bookbindings. Cotton products show up in ice cream and X-ray film, in lacquers and make-up. It is not only jeans that fuel world consumption.

Above A woman's festive cotton dress from Ethiopia. Although *Gossypium herbaceum* enjoyed a long history of exploitation in Ethiopia, most commercial cotton grown there is now from the New World species and hybrids. More drought-tolerant, the local species may provide genetic material for new hybrids for this thirsty crop.

Opposite A plant and seeds of *Gossypium neglectum*, a synonym of *G. arboreum*. Although it is not certain where this Old World 'tree cotton' was first domesticated, it was developed by the Indus valley cultures and the resulting varieties dominated cotton production in Asia before the introduction of New World cottons.

No. 1495

(1493) *Gossypium herbaceum*, W.

G. neglectum Tod.
fs.

Bamboo

Bambusoideae
Versatility and Strength in a Stem

The tip of a bamboo shoot and delicate leaves on a Japanese block of lacquer ware. Bamboo is frequently depicted in East Asian art, both for its symbolism and the overlap with calligraphy techniques.

Labourers are obliged to pare their nails, but people of quality let them grow …
and at night put little cases of bamboo on them.
Peter Osbeck, 1771

When Europeans and North Americans encountered the material culture of the bamboo-rich countries of China and Japan in the 19th century they were astonished by the multiplicity of uses to which the plant's hollow stems were put. Understanding this technology provided a window on to these societies – bamboo substituted for much that was made of wood and metal elsewhere. It could be said that tropical and subtropical Asia was profitably living in a 'Bamboo Age'.

Today there is an increasing realization of the economic potential of the world's fastest growing 'woody' plants (they are not true wood). *Phyllostachys edulis* has been recorded as putting on 120 cm (almost 4 ft) in 24 hours during its maximum growth period, and it can reach up to 28 m (92 ft), yielding stems for the construction industry and fibres for textiles, as well as edible shoots when young.

Bamboos are broadleaved grasses (Poaceae); as forest dwellers, they are the only major grass lineage to have diversified in this environment. There are some 1,400 species (in around 115 genera) identified worldwide. They fall into three tribes: Bambuseae and Arundinarieae are the major ones, while Olyreae live mostly in the Americas. Some bamboo species are difficult to study because of their infrequent flowering. In 1912 specimens of *P. edulis* flowered in Nakasato, Japan. Seeds were planted in Yokohama and Kyoto, and although separated by 350 km (217 miles) the resulting culms all flowered in 1979 (such gregarious behaviour is known as mast flowering), completing their 67-year lifecycle. The estimated time taken by *P. reticulata* from seed to flower is 120 years. Bamboos also reproduce vegetatively by underground rhizomes, which allows their rapid increase. There are both 'runners', which expand horizontally beneath the soil and send up new shoots, and 'clumpers', budding new culms from the sides of the primary plant.

Such diversity of species and bounty of growth are matched by the richness of use, aside from the simply culinary (both as food and the chopsticks to eat it with). Split stems can be woven, and a sheet of bamboo matting can be made into the side of a house or a hat. Baskets might seem commonplace, but the ability to contain and carry is fundamental. Poultry can be taken to market and fish caught in traps; trays hold silkworms. Twisted into cables, bamboo fibres can be combined

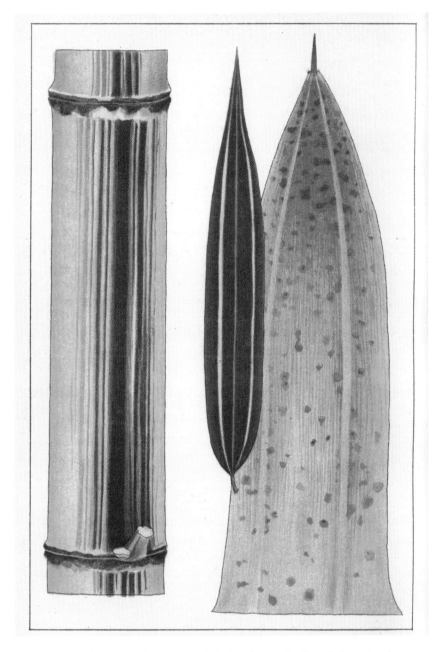

Part of the culm (stem), leaf and sheath of the impressive 5-m (16-ft) tall *Phyllostachys castillonis*. Bamboos are giant grasses of incredible versatility. Construction of China's Great Wall (begun in the 5th century BC) and Grand Canal (5th century AD) both benefited from the bamboo wheelbarrow, which with a sail could carry 165 kg (365 lb).

with whole culms to bridge rivers, while bamboo rafts float below. Bamboo can carry water as well as cross it: as irrigation pipes and water wheels, buckets and cups. A simple pole over one shoulder carries a suspended load on each end, two poles a sedan chair. From outdoor to indoor furniture, the house itself, the temple and the scaffolding for larger buildings, all are made of bamboo. It is used to write with and on – first on split culms and then paper. The giant panda requires none of these, but must simply eat huge quantities of bamboo and have a large enough range to move on when, after flowering, plants of some species die away.

Mahogany
Swietenia spp.
Furniture Timber of Choice

Of all woods, mahogany is the best suited to furniture where strength is demanded. It works up easily, has a beautiful figure and polishes so well that it is an asset to any room.

Thomas Sheraton, 1803

A mahogany tripod table of mid-18th century design – it was extremely fashionable, with its characteristic piecrust rim. Originally, such tables were fashioned from a single piece of wood, but in reproduction pieces the rim is applied afterwards. In fact this is a toy table, but even in a doll's house, only the best would do.

It is easy to see why mahogany has long been favoured as a wood of fine furniture, panelling and woodwork. Its hue (after it dries) is so rich and distinctive that it has become an actual colour, its grain is exquisite, and it is tough and durable.

'True' mahogany belongs to the genus *Swietenia*, named by Nikolaus Jacqui, a Dutch disciple of Linnaeus, in honour of his patron, Gerard van Swieten, professor of medicine in Vienna. Linnaeus had thought it was a kind of cedar. Southern American mahoganies were widespread through the tropical regions of the continent and the Caribbean, and had been long used by people in the area. They consist of three closely related species, *S. mahagoni, S. humilis* and *S. macrophylla.* The first was found predominantly in southern Florida and the Caribbean as far north as the Bahamas; the second, with smaller leaves and preferring drier climates, spread from northern South America into Mexico; the third, the big-leafed favourite of modern plantations, was dominant in Brazil and Honduras. These three species can hybridize freely. They supplied most of the European and North American requirements for beautiful furniture, carvings and building construction. Imports peaked in the late 19th century, by which time supplies were already becoming much scarcer.

Mahogany needs sun and warmth to thrive, and is slow growing and solitary. This means that no mahogany forests exist ready to be exploited: rather, isolated trees have to be individually cut and laboriously dragged to a navigable waterway, road or (now) rail terminus. With proper spacing they can be plantation grown, but the mahogany shoot-borer is a serious pest. Successful plantations have been established in India, Bangladesh, Indonesia and Fiji. In an attempt to preserve this magnificent tree (*S. macrophylla,* can reach up to 70 m/230 ft and 3.5 m/12 ft in diameter), two principal things are being done.

First, mahogany is now listed by the Convention on International Trade in Endangered Species (CITES) and its exploitation is being increasingly regulated. This happened only after massive wastage in the Amazon, where trees were simply felled and burnt when land was cleared for ranch and other food-related use. Despite the restrictions, a large percentage of mahogany trees are still illegally

Swietenia Mahagoni

logged. It is estimated that in Peru, currently the world's largest exporter, up to 80 per cent of mahogany is produced outside international regulations.

Second, 'mahogany' now includes a number of other hardwood trees, some not even in the same family. African mahogany belongs to the genus *Khaya* or *Entandrophragma*; Sri Lankan mahogany is a cedar tree; while Philippine, New Zealand and Chinese mahoganies are yet other types of tree. They are all marketed as 'mahogany'. So today mahogany furniture may or may not be made from the same kind of tree used by Chippendale, the brothers Adam, Hepplewhite or the other brilliant craftsmen and designers of Britain, Europe and the United States during the heyday of the wood. But the enduring qualities of the true mahogany are still appreciated. Apart from furniture, it is used in some of the famous Ludwig drums and for the best electric guitars, including Gibson, because it produces a warm, vibrant sound. The wood's properties of beauty and durability still command respect, and a price to match.

A flowering sprig of *Swietenia mahagoni*, with a five-lobed seed capsule and seed. The capsule persists on the tree through the winter before cracking open to release its seeds to take flight in spring. In its native Caribbean, preparations from the tree's bark were traditionally used as a medicament.

Cash Crops
Making It Pay

Plant products have long been exchanged as barter or for money. Trade networks in the ancient Mediterranean, Asia and the New World were extensive, and often involved timber, food or materials for cloth. Modern exploitation of cash crops differs mainly in its scale and the wholesale transplantation of plants outside their original settings.

Tobacco was moved around the New World before Europeans encountered the 'sot weed', and was used in most of the ways that it still is. Sugar cane had already been transported from the Pacific islands into Asia and Southwest Asia and Africa before Europeans took it to the Caribbean and mainland America. Tea plants were more carefully guarded in China before they, too, were smuggled out, as seeds or young plants, to India, Sri Lanka and eventually to Africa. Coffee gradually spread from its home in Ethiopia, both to the New World and Java, and other parts of Africa. Demand for tea, coffee and sugar grew in parallel, and sugar and tobacco (along with cotton) fuelled the development of slavery in the New World, as plantation growers expanded production of these labour-intensive crops.

Like coffee and tea, chocolate also contains stimulants; unlike its original New World connoisseurs, Europeans preferred it sweet. They also adapted its processing and vastly extended production beyond its Central and South American homeland to several tropical areas of Africa, from where much of it now derives.

The Flower, fruit, & plant of the Bonanas.

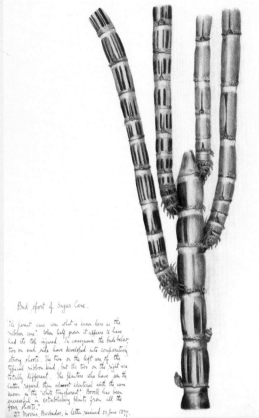

Bud sport of Sugar Cane.

The parent cane was what is known here as the "ribbon cane" when half grown it appears to have had its top injured. In consequence the buds below two on each side have developed into comparatively strong shoots. The two on the left are of the typical ribbon kind, but the two on the right are totally different. The planters who have seen the latter regard them almost identical with the cane known as the "white transparent". Bovell has been successful in establishing plants from all the four shoots.
Dr Morris, Barbados, in letter received 22 June 1899.

Africa in turn donated the oil palm to Malaysia, Indonesia and Thailand, as well as to tropical Central and South America. The development of large plantations in areas the palm was introduced to has inevitably resulted in environmental damage, with loss of local habitat and rainforests, together with the issues that monoculture always brings.

The banana, now grown far beyond its home in Southeast Asia and available all year round, does not produce seeds, which means that plants are clones with no natural genetic variation, a worrying addition to the usual pitfalls of monoculture. Plants were taken to the New World to provide local food for slaves, and refrigerated shipping from the late 19th century helped expand the market.

The unusual properties of rubber's sap were known by people in the Amazon, but its successful commercial exploitation needed the input of chemical technology (vulcanization) to stabilize the product and make it more resilient to temperature. First bicycles, and, above all, automobiles, vastly increased the market for it and led to massive rubber plantations in parts of Africa, China and the Philippines. Chemistry, too, has dramatically affected world supplies of indigo, long prized for its rich blue colour, whether from the Asian indigo tree or the European herbaceous woad. Valued for thousands of years, indigo is now mostly synthesized in the laboratory.

Above left A flowering, fruiting stem of a banana, with a banana tree to the left, in a detail from a vivid hand-coloured plate in John Cowell's *The Curious and Profitable Gardener* (1730). A note in the Kew Library copy, from the library of Sir Hans Sloane, comments that these were 'painted in their natural colours'.

Above right 'Bud sport of Sugar Cane', an original artwork from a set recording various trials and experiments on Barbados at the end of the 19th century. These sweet grasses come in a wonderful range of colours and patterns.

Tea

Camellia sinensis

Tips of a Global Trade

I am in no way interested in immortality
But only in the taste of tea.

Lu Tung, 9th century

Tibetan teapot from Shigatzi, brought to Kew by Joseph Dalton Hooker after his forays in search of plants. Traditionally, Tibetans favoured brick tea, which is first steamed, compressed and aged. Broken up again, the tea is then churned with salt, yak butter and hot water in a bamboo vessel, and kept warm in the teapot on a brazier.

Both the tea plant, *Camellia sinensis,* and tea drinking originated in China. The plant is an attractive, evergreen shrub (or tree if allowed to grow), which needs elevation, warmth, good rainfall and an acid soil. Like coffee and certain other plants, tea leaves contain two especially important alkaloids: caffeine and theophylline. Both are stimulants and both are addictive, which helps explain why tea (and coffee) drinking has long been so widespread.

The origins of tea drinking in China are shrouded in myth, but tea leaves were gathered and used in infusions possibly by the mid-1st millennium BC. The plant's leaf was probably chewed even earlier, as it still is today in the southwestern regions of China where *C. sinensis* is indigenous. Tea's cultivation, preparation and use are intimately intertwined with China's ancient and turbulent history. It was used as a currency and an official means of payment; it was frequently taxed at a high rate and subject to a state monopoly to maximize revenues. Its tastes were revered and it achieved an almost religious significance.

The two major kinds of tea, green and black, are simply the result of different picking and processing. Only the tips – the top two leaves and a bud – are picked, and this is still done by hand. Green tea is slightly younger when picked and not 'fermented' (oxidized) so much. More delicate, it was the preferred brew in China. Black tea is processed more fully. It is more robust and became the favourite in many nations abroad. Between these two, oolong is also popular. The 'curing' of tea is an elaborate process, requiring great knowledge and skill. Properly prepared and packed, tea keeps well, an important factor in its eventual worldwide popularity.

Tea drinking was so fundamental in Chinese society that ceremonies evolved around it. Such ceremonies were even more ritualized in Japan, after Buddhist monks studying in China took tea back to their native country in 805. The Japanese tea ceremony reached its highest form with Sen Rikyū in the late 16th century. It was bound by strict rules and took place in a precisely constructed room with a host and five guests. The entrances and exits, implements, dialogue and order of events were carefully orchestrated. In a sense, the tea itself was almost incidental to the wider meanings of the ceremony, but it had to be properly prepared, brewed and served to perfection. Variants of the ceremony were introduced after Sen's execution in 1591 (for reasons that are unclear), but the ceremonial importance of tea has remained a feature of Japanese life.

Tea was central to Chinese society too; and Mongol invaders also developed a taste for the beverage, mixing it with milk and butter. Tea also gradually spread overland into Russia and other Asian countries. The first Russians recorded to have tasted it were two envoys sent in the early 17th century to negotiate with a Mongol prince. After the drink had become popular throughout the vast country, the samovar came to embody Russian domesticity, with its clever internal pipework keeping hot water and tea always on tap. Most tea reached Russia via Kyakhta, a once prosperous town on the Mongolian border – its tea market was the reason for its historical status.

Tea arrived in Europe by boat. After the opening of the ocean route to Asia via the Cape of Good Hope in the late 15th century, reports of the beverage began to trickle back to Portugal and the rest of Europe. The Dutch began importing it to Europe in the early 17th century, and the diarist Samuel Pepys recorded his first experience with the 'China drink' on 25 September 1660. Tea had just become available, and was sold at Thomas Garway's famous coffee house in Exchange Alley, London, and along with coffee it quickly became fashionable. Import taxes meant that much of the tea was smuggled.

For almost two centuries, Europeans did not know where in China the plant was grown, since trade was carefully controlled and few foreigners were allowed beyond Canton, the main port. In fact, the transport of the tea from the inland sites of growing and processing involved heroic feats of organization and physical labour. Understandably, it was assumed at the time that green and black tea came from different tea plants.

Transporting tea by raft, packing and labelling tea crates, and weighing tea – a Chinese business in Chinese hands. By the 19th century, much of China's official trade passed through the port of Canton, where foreign merchants could set up trading houses or factories but were not allowed inland. They were obliged to deal with the government-designated merchants or 'Gong Hang'.

Above left 'Bergamotto Foetifero da Padoua' by Johann Christoph Volkamer in *Nürnbergische Hesperides,* (1714). The essential oil from the rind of the Bergamot orange (*Citrus bergamia*) gives the unique flavour to the tea blend known as Earl Grey. This orange is thought to be a hybrid between the sweeter *C. limetta* and the bitter orange, *C. aurantium.*

Above right An idyllic scene of a tea plantation from Samuel Ball's *An Account of the Cultivation and Manufacture of Tea in China* (1848). Ball served as inspector of teas for the East India Company in Canton and wrote his book after returning to Britain, to assist those trying to cultivate tea in British India and other parts of the empire.

The benefits and dangers of tea drinking were long debated, and the 18th-century lexicographer Samuel Johnson, an inveterate consumer, had to defend his habit. Gradually, however, the beverage installed itself at the centre of British life. British demand for tea, and having to meet the Chinese preference for silver to pay for it, led to a desire to find an export to China to balance expenditure. Opium fitted the bill, and the 19th-century Opium Wars were partially fuelled by the British obsession for tea. In the mid-19th century the 'tea clippers' competed to be the quickest from Chinese ports to English ones, and the races were deemed newsworthy, although properly prepared and packed tea stores well, so it was not really about the quality of the final drink.

Consumer demand also encouraged the search for other areas for tea growing, and after early difficulties cultivation of the plant enjoyed increasing success in British India, first in Assam and then in Darjeeling and other hillside areas. This encouraged tea plantations elsewhere, including Kenya and other parts of East Africa. In Ceylon (Sri Lanka), tea proved a blessing after the coffee plantations were wiped out by a fungal disease.

Tea continues to be one of the most popular beverages in the world, especially in Britain, Australia and its areas of early production, China and India. In the modern world, 'brands' have come to dominate. These often still bear the names of previous entrepreneurs, such as Lipton and Twining; Earl Grey tea, with added bergamot flavourings, also bears a famous name. The tea bag, introduced around 1908, and derided by connoisseurs, finally democratized this once elite drink.

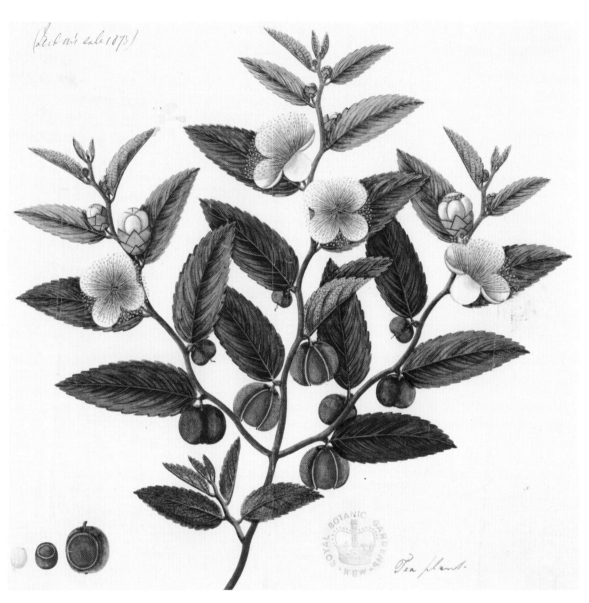

A flowering sprig of the tea plant, *Camellia sinensis*, in the same genus as garden camellias. The tips or terminal flush, 'two leaves and a bud', are picked twice a year in early spring and late spring/ early summer. The 'bud' is not an unopened flower but an immature unopened leaf.

Coffee
Coffea spp.
Waking Up the World

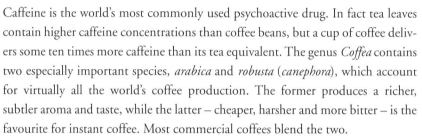

Coffee makes us severe, and grave, and philosophical.

Jonathan Swift, 1722

Opposite Coffea arabica from a multi-volume work by Dr Friedrich Gottlob Hayne (1825). Despite its easy availability as a beverage, coffee was still regarded as a medicinal substance. Relatively pure caffeine was isolated in 1819 by Friedlieb Ferdinand Runge.

Below A Turkish gentleman sips his coffee, above, with the plant, beans and a coffee roaster depicted below, in this delightful engraving from Philippe Sylvestre Dufour's *Traitez nouveaux & curieux du café, du thé et du chocolate* (1688).

Caffeine is the world's most commonly used psychoactive drug. In fact tea leaves contain higher caffeine concentrations than coffee beans, but a cup of coffee delivers some ten times more caffeine than its tea equivalent. The genus *Coffea* contains two especially important species, *arabica* and *robusta* (*canephora*), which account for virtually all the world's coffee production. The former produces a richer, subtler aroma and taste, while the latter – cheaper, harsher and more bitter – is the favourite for instant coffee. Most commercial coffees blend the two.

Coffee belongs to the tropical family Rubiaceae and the bush grows best at altitude in warm climates with plenty of rain. It is native to present-day Ethiopia, where it is still cultivated but also grows wild. According to legend, a young goatherd noticed that his goats that had eaten the coffee berries (the seeds, or beans) were frisky and unusually lively. He subsequently got hooked on them himself. Legend aside, sucking coffee beans was almost certainly the earliest method of use. Eventually, the leaves and berries were boiled to yield a drink containing the stimulant caffeine.

Medieval Arabic texts first mention the beverage, and for several centuries its use was confined to northern Africa and southwestern Asia. Coffee became a favourite with Muslim clerics as it enabled them to continue their meditations through the night. The beans were later transported east, where they were cultivated in India, Sri Lanka and other Asian countries by the early 17th century, about the time that Europeans discovered what a cup of coffee could do. There was a coffee house in Constantinople (Istanbul) by about 1475, and Europeans there enthused about the new beverage.

The Coffee House became a central part of European culture from the mid-17th century in cities such as London, Paris, Amsterdam, Vienna and Berlin. In these establishments the availability of newspapers and the discussions on a range of current affairs were as important as the beverage, but coffee, along with tea – both drinks appeared roughly simultaneously (as did chocolate) – were the reason the place was there. In Vienna, coffee houses continued to serve important cultural and political functions into the modern period. The Viennese had discovered

Coffea arabica

the bean after the siege of the city was lifted in 1683 and the retreating Ottoman army left behind their supplies, including the precious beans.

Coffee has often vied with tea as a nation's favoured caffeinated drink, and both have served the same sociable and stimulating roles. The French flirted with tea, but coffee won out in the 18th century. Paris had 2,000 coffee houses on the eve of the French Revolution. In Britain, tea supplanted coffee in the 19th century: it was cheaper and easier to transport and store. In the United States, where there were coffee houses as early as 1670, coffee was the victor, and the modern history of coffee is closely tied to the economic and political fortunes of that country, which in the 19th century was by far the leading coffee consumer in the world. By this time coffee cultivation had spread to many parts of the globe where conditions and labour supply were favourable.

Generally these regions were part of European colonial possessions, and included Brazil, Colombia and Costa Rica in the Americas, Kenya in Africa, and (briefly) Ceylon (Sri Lanka) and (more lastingly) Java in Asia. Brazil has long been a leading coffee producer. The Brazilian harvest has periodically been affected by adverse weather or occasional drought, but mostly by sharp cold, which the plants do not tolerate. Coffee was so important for Brazil's balance of payments that successive governments aided the growers' associations, helping stockpile gluts to keep the international price up, and sometimes tiding over growers in times of poor yields. International financiers, mostly American, also invested in coffee planters and stores.

Cultivation of coffee is labour intensive, since the beans must be gathered by hand – they ripen sequentially so mechanical means don't work. The raw coffee beans are generally exported for roasting in the country of consumption. Roasting

The flowers (right) and the subsequent beans (left) of *Coffea arabica*. This watercolour is by Manu Lal (fl. *c.* 1798–1811), painted in the Company School style developed by Indian artists for the British to meet their need for naturalistic representation of plants, animals and people. Many artists, including Lal, worked for the East India companies that ran the country.

Plantations grew coffee in bulk and also processed it. Here, coffee is being dried by steam in Brazil in an illustration from Francis Thurber's *Coffee: from Plantation to Cup* (1881), published by the American Grocer Publishing Association. Thurber travelled widely and reported on many aspects of coffee production and consumption, including the poor working conditions on many plantations.

is easy in theory, but hard to get right. The longer the pale beans are roasted, the darker the bean will become. The process releases the aromatic and volatile compounds that give coffee its aroma (one of the best in the world) and taste. The aroma is there, briefly, in a recently opened jar of instant coffee, although the taste is different. A Belgian named George Washington developed an instant brew in 1906, in Guatemala. (He was not the first.) Washington subsequently emigrated to the United States. The First World War proved a godsend for the new product, which could be easily made wherever almost boiling water could be produced. American troops called Washington the 'soldier's friend'. Instant coffee is made now mostly by freeze-drying brewed coffee, which produces the powder or granules in the commercial jar.

For many people today, 'coffee' is instant coffee, but the rise of the new wave of 'coffee houses', which cater for a computer generation as the old ones did for those in search of a newspaper and warm fire, has produced its own modern clientele, who care about the source, taste and style of their caffeine drink.

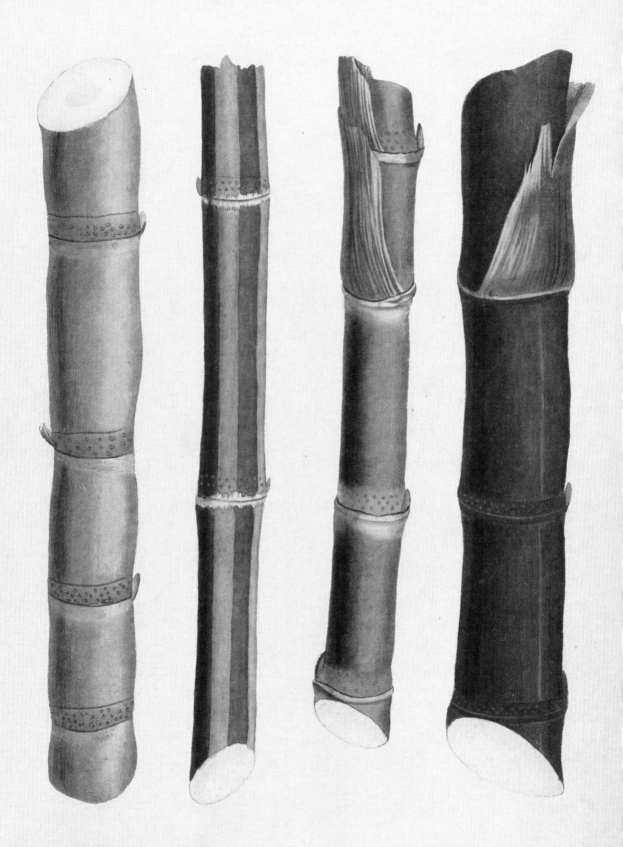

Sugar Cane

Saccharum officinarum

The Slave Trade's Sweetener

Around thee have I girt a zone of sugar cane to banish hate.
That thou mayst be in love with me, my darling never to depart.

Hymns of the Atharva Veda

A sweet tooth is nothing new. Long before the ready availability of cheap, highly refined sugar, human beings sought out naturally occurring sweet tastes; sugars are ubiquitous in plants and animals. An especially rich source of sucrose, the 'sugar' of sugar, is *Saccharum officinarum*, or sugar cane. A member of the grass family, sugar cane was probably first cultivated in New Guinea, though the genus is another that hybridizes easily and the wild original species is unknown. Unlike many canes, such as bamboo, which are hollow, the centres of the sugar cane stalks are fibrous and the juice contains about 17 per cent sucrose. This happy fact was discovered long ago: sugar cane was chewed in the Pacific islands and grown in India – Alexander the Great sent sugar cane home from there – and China well before the Christian era. Strabo, writing in the early 1st century AD, repeats earlier accounts of 'reeds that yield honey' without bees.

The process of extracting and boiling down the sap to make solid sugar was discovered in antiquity; the Indians called this raw brown solid 'jaggery'. By the 7th century AD the Persians had discovered that adding lime could render the product white. The Arabs established sugar plantations throughout Southwest Asia, North Africa and on several islands in the Mediterranean, usually with irrigation. From there, sugar gradually made its way into Europe, although it was very expensive; honey, in which fructose is the primary sugar, was still the main sweetener used.

Sugar canes are tropical plants that grow well on rich soil with plenty of rain and sunshine; some can reach up to 6 m (almost 20 ft) tall. As the European taste for the luxury item increased, the Iberian countries established plantations on Madeira and the Canary Islands. Columbus took cane with him on his second voyage to the New World, where the Caribbean isles proved suitable for its cultivation. Sugar was expensive because it was laborious to produce. The tough canes had to be cut near the roots, the leaves stripped off and then quickly crushed and the juice boiled down, filtered and boiled down again. Planting and harvesting were backbreaking work. The cane grows best if a portion containing a node is planted, either in a trench or a single hole. The fields must be weeded and watered

Opposite Four different coloured varieties of sugar cane as featured in François Richard de Tussac's flora of indigenous and imported exotics, *Flore des Antilles* (1808).

Below A bottle of crude sugar brought back from David Livingstone's second Zambezi expedition by Sir John Kirk, who acted as the official doctor and naturalist in 1858–63. A keen botanist, highly regarded by the directors of Kew, he became the Vice-consul of Zanzibar and successfully brought an end to slavery on the island with the co-operation of the sultan.

SUGAR,
Saccharum officinarum L.
TETTE.
Dr. Livingstone's Exped.

if rain is insufficient, and the processing, no easy task, used a lot of fuel. Madeira was largely deforested for cane growing and production.

With the coming of tea, coffee and chocolate to Europe in the 17th century the demand for sugar increased markedly. Manpower was needed to work the plantations in Barbados, Jamaica, Brazil and other New World locations. Indentured labour schemes proved inadequate, and African slavery was the terrible solution. Between 1662 and 1807, when Britain's trade was abolished, some 3 million Africans were transported in her unsanitary and overcrowded ships. The Portuguese, Spanish, Dutch and Americans also had active slave transportation systems for sugar plantations in Brazil and other colonial possessions, as well as Louisiana in the United States. The journey from Africa to the Americas, known as the 'Middle Passage', was part of a triangular route – ships left Liverpool, London or Bristol laden with merchandise to be exchanged for slaves on the west coast of Africa; after transporting them across the Atlantic to the plantations, the ships would return with the prized cargoes. African slavers captured their unfortunate victims, mostly men but also women and children, from the interior of the continent. Survival rates were variable on the long crossing, but conditions were routinely atrocious, with little and bad food, and crowded, inhuman conditions.

The nature of sugar cane cultivation made plantation settings desirable. They varied greatly in size, and were often referred to not by their area but by the number of slaves. Slavery helped define the histories of the New World areas where it was practised, and beyond. After the abolition of slavery in the 19th century, by the French, British, Americans, Dutch, Spanish and Portuguese, the importation of sugar workers from Asia, the Polynesian islands and the Mediterranean countries left an indelible mix on the ethnic compositions of the sugar-producing lands.

It has been estimated that it took one African life to produce a tonne of sugar, but such was the scale of the operation that the price of sugar in Europe and North America came down dramatically from the late 17th century, and it became a commodity within reach of most people. The final refining process was generally done outside the areas of supply, since fuel and power were more easily available there. The British even re-exported some sugar back to the producing colonies.

The basic cultivation requirements for growing cane remain the same, but the production process has changed over the past century. Cutting the cane is now done by machine, making level fields preferable, and pressing and extracting the juice are also mostly mechanized. Little is wasted: the residue of the first boiling, molasses, is used in cooking or distilled to make rum. Sugar distillation yields alcohols and biofuels, and other products include chemicals, fertilizers and animal feeds. Today, the major producers of sugar cane are Brazil and India.

A rival of the cane and another source of table sugar is sugar beet (*Beta vulgaris*). A German chemist, Andreas Sigismund Margraf, discovered in the middle

In *Food-grains of India* (1886) the agricultural chemist Arthur H. Church described the sugar cane as ready to cut when about to produce the 'large feathery plume of flowers', and included details of these flowers in the plate which illustrated *Saccharum officinarum*. Once the cane has flowered the sugar-containing stalk stops growing.

of the 18th century that significant concentrations of sucrose could be extracted from this root crop. Sugar beet can tolerate a wide range of soils and climates, and during the British blockades of the Napoleonic Wars, Napoleon encouraged its cultivation and processing in France, to keep his soldiers sweet. The end table product, sucrose, is the same as from cane, and today about 20 per cent of sugar worldwide is derived from the beet. Despite the global epidemic of obesity, diabetes and other sugar-related 'diseases of civilization', sugar remains very popular. Its sweetness has brought pleasure but also pain.

Chocolate

Theobroma cacao

Food of the Gods

Up, and Mr. Creede brought a pot of chocloatt ready made for our morning draught.

Samuel Pepys, *Diary*, 6 January 1663

For many of the ancient cultures of Mesoamerica and South America, the chocolate tree held a special place. It featured in the creation myths of the Maya, and the Aztecs used the beans (seeds) as a currency. It was so prized by the Aztecs that they imported it over long distances. The even older Olmec civilization also valued the beans of the tree, and 'cacao' probably derives ultimately from their language, now unfortunately lost. The Maya referred to the plant, its seeds and the product as cacao, and cognates exist in other Mesoamerican languages. Linnaeus, who was fond of chocolate, bestowed its current botanical name: *Theobroma*, meaning 'food of the gods', and the native word *cacao*.

T. cacao is a fastidious tree that doesn't naturally grow beyond a range of about 20 degrees north or south of the Equator. It needs shade, high temperatures and humidity. On modern plantations, the over-storey is generally provided by either rubber or banana plants. The pods grow directly from the tree's trunk and stems (a phenomenon called 'cauliflory'), and the flowers are fertilized exclusively by midges (they do have some use). Only a tiny fraction of the flowers go on to produce a pod, and a good tree will produce about 30 pods each year. Within the pod, the pulp surrounding the seeds is sweet but the raw kernels are quite bitter. They must be fermented, dried, roasted and winnowed (the removal of the thin shell) before the cacao 'liquor' is produced. Debate continues about where the tree originated – probably Amazonia – but it was domesticated in Mesoamerica. Another species of *Theobroma* (*bicolor*) is also grown from southern Mexico to Brazil, where its product, called *pataxte*, is drunk or mixed with the seeds of the more expensive *cacao*.

Native Americans ate the juicy pulp or drank the ground beans in a beverage, mixed with a variety of flavourings including chillies and vanilla. As befitted a substance so venerated, it was used on ceremonial occasions. The Maya had a cacao god, with a regular festival. Elaborate utensils for drinking chocolate survive, and the vessels, and probably the beans themselves, were left in the tombs of important individuals. One cache, thought to be actual beans that had miraculously survived the heat and humidity of Mesoamerica, turned out to be models beautifully made

THEOBROMA CACAO

Varieties of raw cacao kernels from chocolate's homeland and places around the world where it was transplanted: Ceylon (Sri Lanka), Guayaquil, Caracas, Portuguese Africa, Trinidad, Samoa. These boxes were once part of an exhibit at the Museum of Materia Medica at University College Liverpool (later the University) – a reminder of chocolate's long association with medical use and the continued claims for the health benefits of dark chocolate's flavanols.

from clay in the shape of the bean. Always expensive, the beans were reserved for elites and the rich. Chocolate was thought to be intoxicating and too dangerous for women and children. The beans do contain a complex set of alkaloids, including caffeine and theobromine, that today would be described as stimulants rather than intoxicants.

Columbus encountered the beans in a captured canoe during his third voyage to the New World, but it was not until the Spanish landed in Mexico that Europeans sampled the exotic drink. They didn't take to it immediately, but soon learnt to flavour it with vanilla and other spices. Sugar was eventually added to make it the sweet drink it now is. The beans reached Spain by 1544, and were a commodity rather than simply a novelty by 1585, although as in their homeland they remained so expensive that only royals and elites could indulge in them. Chocolate gradually spread to other parts of Europe, including Italy. The French discovered it in the early 17th century, and by 1657 there was a chocolate seller in London, with the drink soon available at the tea and coffee houses. Europeans generally preferred to take it sweetened and hot, although the Spanish continued to add chillies.

Once the commodity was more widely available, cooks began to incorporate it too, although it was mostly used either medicinally or socially. Increasing European demand led to further planting of the tree in several Caribbean islands, including Trinidad and Jamaica. When the British captured Jamaica from the Spanish in 1655, it became a major supplier for the British market. The plantations had grown the original Mesoamerican variety, Criollo, which produces fine chocolate but is very disease prone. After a blight virtually wiped out the Trinidad plantations, a more robust variety, Forastero, replaced it. Forastero had been discovered growing wild in Brazil. It now accounts for almost 80 per cent of world production, although hybrids are being developed that combine the robustness of Forastero with the finer flavour of Criollo. The international market has resulted in the spread of *Theobroma* cultivation to many areas within the climatic constraints of the plant; West Africa is now the world's leading producer.

The processing of the pods yields a number of different products, each of which has uses. The sun fermentation of seed and pulp is essential to develop the flavours and results in a full-fat liquor, which was what was consumed until the early 19th century. Then, in 1828, a Dutchman, Coenraad van Houten, with his father, patented a process that reduced the paste by about two-thirds, yielding a product called cocoa. A portion of the cocoa butter that had been extracted could then be added back to the residue, producing the solid (which melts in the mouth) that we call chocolate. Within a couple of decades, chocolate bars were on the market.

Many of the familiar names in chocolate confectionery date from the 19th century: Cadbury in Britain, Lindt in Switzerland, Hershey in the USA. As with any mass-produced product, the finished product varies and is generally determined by the percentage of cocoa solids. 'Milk chocolate' was invented when a Swiss confectioner, working with Henri Nestlé, added dried milk in 1876. There is now a chocolate for every taste.

Tab. 99.

NICOTIANA TABACUM. L.
Der gemeine Tobak.

Tobacco
Nicotiana spp.
The Sot Weed Factor

A custome loathsome to the eye, hatefull to the nose, harmfull to the braine,
dangerous to the Lungs.

King James I of England, 1604

Tobacco has been almost universally demonized, but it is still much in demand. This is not just a modern phenomenon: ever since its introduction to Europe in the early 16th century, there have been critics as well as advocates.

Although species of *Nicotiana* existed in Australia, the current tobacco plant originated in the New World, where it is probable that it was deliberately cultivated several thousand years ago. Of two native species, *N. rustica* and *N. tabacum*, the latter provides most of present-day production. Tobacco's original home was likely the eastern parts of South America, but varieties were widespread through the western hemisphere in pre-Columbian times, as was use of the plant. A member of the Solanaceae family (which includes the tomato and potato), tobacco is an annual that grows to a height of between about 20 cm and 3 m (8 in. to 10 ft), depending on the variety and growing conditions. It is grown commercially for its leaves, which are harvested and then dried for use.

Tobacco leaves contain numerous alkaloids, of which nicotine, isolated by two Germans, Wilhelm Heinrich Posselt and Karl Ludwig Reinmann, in 1828, is by far the most potent. Tobacco is both psychologically and physiologically addicting. In high concentrations it can also be hallucinogenic, a fact appreciated by priests and shamans of pre-Columbian America. Both the Maya and the Aztecs prized the leaf. American peoples used the 'sot weed', an early European appellation for the substance, in many of the ways that it still is – they made cigars, smoked it in pipes, took snuff and used it in teas and enemas. Tobacco was also important ceremonially, with North American tribes creating elaborate pipes, some for war and some for peace.

Columbus was given tobacco on his first voyage. Although his men initially discarded the leaves, on the second landing, in modern-day Cuba, some of the sailors tried smoking, and tobacco was among the New World products taken back to Spain. At first tobacco was also used medicinally, but the smoking habit spread fairly quickly to the Mediterranean countries, and to Britain, where tobacco was introduced around 1564. Queen Elizabeth I tried it, but her successor, King James I, was violently opposed to the weed. He did, however, appreciate the boost to his

Opposite Both the plant, *Nicotiana tabacum*, and (later) the alkaloid were named for Jean Nicot, who served as French ambassador in Lisbon, Portugal, in the 16th century. Here he learnt about tobacco, particularly its medicinal use, and sent seeds back to the French court where it became popular with Queen Catherine. Instead of smoking the leaves, he advised her to crush them and inhale, like taking snuff.

Below An Indian nabob rides on his palanquin, smoking from a hookah or hubble-bubble pipe carried by one of his servants. In India tobacco might be mixed with sugar and rosewater. In Egypt and elsewhere it was blended with fruit, mint and molasses to produce shisha, hence the similar shisha pipe. In both, the smoke is bubbled through water before being inhaled.

purse that the duties (which he increased dramatically) provided, though they also encouraged smuggling.

In Europe most tobacco was initially smoked in pipes, with some also rolled into cigars. In the 18th century, taking snuff (ground tobacco sniffed up the nose) grew in popularity, a custom accompanied by the manufacture of ornate snuff-boxes. Tobacco was officially a male preserve, although ladies began to take snuff and lower-class female tobacco smoking also increased. The habit even influenced domestic architecture and the rituals of dinner, with the ladies withdrawing after meals to leave the men to their pipes, cigars and port. Large houses had their smoking rooms, sometimes with a billiards table.

Tobacco was the first major cash crop for the original English settlements in Virginia and played a large role in their financial viability. As early as 1613, the leaf was being exported to England, and by the mid-18th century it was the major revenue earner for the colonies on the southern eastern seaboard. At first tobacco growing was small-scale, but as tenant farmers strove to satisfy demand, operations expanded by using African slaves. Tobacco plantations were also established on several of the Caribbean islands.

Tobacco growing spread to the American Midwest, and in the 1860s an Ohio planter noticed a new variety that was to change the face of the industry. Cigarettes, with the tobacco wrapped in paper, had been smoked in some form even by pre-Columbian Americans, and British soldiers in the Crimean War (1853–56) had learnt the custom from Turkish use. The new tobacco, called burley, has little chlorophyll and its yellow leaves are much milder. Burley became the favourite for machine-rolled cigarettes since it was much easier to inhale and smokers absorbed more nicotine, with addictive results.

Two occurrences in particular facilitated the meteoric rise of cigarette smoking around the turn of the 20th century. The first was the increased power of advertising following the arrival of mass-print newspapers and magazines in the mid-19th century. As large multinational tobacco companies sought to capture brand loyalty, the advertising was ubiquitous and clever, targeting different groups, such as young people, doctors and women. The last group massively expanded the market, as smoking became associated with the 'New Woman', who was far more independent and had money to spend. The second strand was the advent of the First World War, which made cheap, often free, cigarettes available to millions of young people on all sides of the conflict, and behind the lines as well. Many went on to smoke regularly after the war, and the pattern was repeated in the Second World War.

In the first half of the 20th century, smoking was widespread, glamorized and affordable for almost everyone, even though taxed by governments, who welcomed the revenues. At the same time, doctors began to notice an increase in lung

cancer, a rare disease before the beginning of the century. Almost simultaneously, two pairs of researchers in the early 1950s, one in the USA and one in Britain, published careful studies that implicated cigarette smoking in the modern epidemic. The American pair, Ernst Wynder and Evarts A. Graham, correlated smoking statistics with autopsies on patients dying from the disease. The British pair, Austin Bradford Hill and Richard Doll, who thought at first they might be looking at a consequence of the material used in paving roads, did more careful life histories of patients in hospitals with lung cancer. They factored out many possible causes and concluded that cigarette smoking was the culprit.

In retrospect, the evidence was conclusive, but it took more than a decade for it to be widely accepted, and tobacco manufacturers have been remarkably devious. First denial, then 'safer' options and then the right of choice have been very effective, and, even in countries where tobacco advertising has been banned, smoking rates have come down only gradually. In developing countries they are still very high, and the sot weed seems destined to be around for a while yet.

Pipe designs from *Il Tabacco Opera* (1669). The use of tobacco and its products quickly gave rise to a whole range of paraphernalia of containers and equipment, some becoming beautiful objects in their own right.

Indigo, Woad
Indigofera tinctoria, Isatis tinctoria
Searching for True Blue

We are not furnished but two drugs that give a permanent blue,
and they are, indigo and woad.

Elijah Bemiss, 1815

Indigofera tinctoria was grown as a plantation crop in the 18th century in South Carolina, Florida, the Caribbean and in South America, where John G. Stedman witnessed its production and recorded it in his *Narrative of a five years' expedition against the revolted Negroes of Surinam* (1796). He was appalled by the brutality of the slave owners.

Blue is a rare colour in the plant kingdom and even rarer in the foods we eat, but it has always been prized. It is the colour of mourning in some countries, the robes of the Tuareg in North Africa and warriors in ancient Britain. Most of this blue colouring was laboriously extracted from plants, two of which were especially significant: *Indigofera tinctoria* ('indigo') and *Isatis tinctoria* (woad). The word indigo refers to the dye extracted from the leaves of these and a number of other plants and derives from 'India', where much of the precious dye originated.

Producing indigo is a challenging and complicated process, but one that was discovered independently many times, in both the pre-Columbian New World and in the Old. The leaves must be mixed with lime or stale urine and allowed to ferment in vats. After the liquid is evaporated and further manipulated, the resulting brilliant blue powder could be easily caked and transported. It had many uses, in dyes, paints and cosmetics, and provided a base to mix with other colours. Not surprisingly, so valuable a substance was also used medicinally against many disorders, including cholera and bleeding maladies and as an aid to fertility.

Indigofera tinctoria is a tropical shrub-like plant, especially common in India, but also found throughout Southeast Asia. It was cultivated and exploited in its native setting from the end of the 1st millennium BC. Indian textiles were valued from an early period and indigo was traded via the Silk Roads and other routes. Both the ancient Greeks and Romans prized it, but they also had to rely on woad, which is a herbaceous biennial (of the brassica family) that grows in temperate climates. It yielded the same valuable dye and was closer to hand, but extraction was more difficult, since concentrations of indigotine (the chemical name for the blue pigment) are much lower. There is still debate about how ancient Egypt got its blue dyes – it could have been from woad, but *Indigofera* was traded in the area very early, and the Muslims continued the fascination with the colour.

European shipping routes to Asia made proper indigo more readily available, and after the British established hegemony in India they exploited both its importation into Europe and its use in their own expanding garment industries. European woad production saw a revival when Napoleon placed an embargo on British

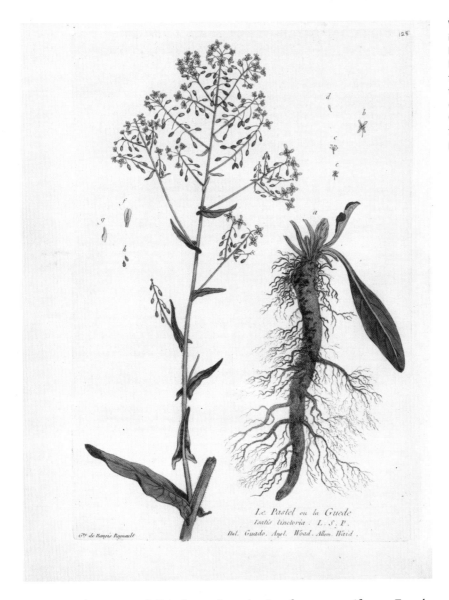

Le Pastel ou la Guede
Isatis tinctoria . L. S. P.
Ital. Guado. Angl. Woad. Allem. Waid.

Woad provided the blue increasingly sought after in Europe from the 12th century. It was widely associated with the Virgin Mary and became the chosen colour of French court. The town of Erfurt in Germany built its university on the proceeds of growing and processing woad.

imports and encouraged French woad production for army uniforms. French, German and some British woad growing continued throughout the 19th century. Towards the century's close, two events had a great effect on modern indigo.

The patenting of metal studding on blue jeans by Jacob Davis and Levi Strauss & Co. in 1873 gave them the necessary durability required by ranchers, farmers and cowboys. The universal appeal of jeans has ensured a continuous demand for indigo, whose characteristic uneven fading has become a fashion statement. This need has been met by the efficient synthesis of indigo, by Adolf von Baeyer, in 1897. Most natural indigo production is now small-scale, with some in India and Africa, and some by artisans in the West who still value the true blue that nature yields.

Rubber

Hevea brasiliensis

Amazonia's Precious Latex

... the pith of a wood that was very light.

Andrea Navigero, 1525

Opposite Sprig of *Hevea brasiliensis*, with flowers (lacking petals and described as pungent) and the three-seeded nut. When ripe, the fruit explodes and the seeds can be thrown up 15 m (50 ft) from the parent tree. In plantations trees grow 25 m (82 ft) high because of the effects of tapping and have a useful lifetime of perhaps 35 years. Rubberwood is now marketed as an eco-timber rather than burned as waste.

A number of plants produce rubber, but only the Brazilian 'rubber plant' is employed commercially today. It is one of many gifts of the New World to the Old, and it was exploited and revered long before the Europeans came. The Amazonians knew of the striking waterproofing properties of the latex sap that can be tapped from the tree, and rubber use was widespread in pre-Columbian Mesoamerica and South America. Containers, shoes and musical instruments were made from the hardened substance, which could be moulded as it dried, and it also had medicinal, ritual and ceremonial applications.

In the early 16th century Aztec teams demonstrated a game using a rubber ball (made from another rubber-producing plant) to the court in Spain. Indeed, Europeans seemed most fascinated by the ball game, the object of which was to put the rubber ball through a hoop without using the hands or allowing it ever to touch the ground. Only in the 18th century did naturalists take a serious interest in the tree and its product. A French engineer and amateur botanist, François Fresneau, described both the tree and the tapping for the sap in 1747, and others in South America brought home accounts of the unusual properties of rubber.

Early European attempts to exploit the substance were not successful. Various entrepreneurs in the early 19th century imported rubber to make boots and raincoats and waterproof cloth. (It was also, more successfully, noted that the material could remove pencil marks: the 'rubber', or eraser, had arrived.) All these enterprises failed, because in very hot weather the rubber melted, and in very cold it hardened and cracked.

The breakthrough came in 1839, when an eccentric American, Charles Goodyear, discovered after a great deal of empirical experimentation that adding sulphur to the melting rubber stabilized it; extreme temperatures no longer had their undesirable consequences. Goodyear spent much of his life moving his family from place to place, seeking backers and spending some time in debtors' prisons. Although his name was posthumously perpetuated in a major international tyre company, he didn't reap the rewards of his successful discovery. At about the same time an English chemist, Thomas Hancock, studied the process in more detail,

Hevea brasiliensis Müll. Arg.

with better chemical knowledge, and obtained an English patent for vulcanized rubber (he called it 'ebonite', but Goodyear's 'vulcanization' has lasted).

Goodyear never stopped preaching the multiple uses of rubber. At the Great Exhibition in London (1851) and the Exposition Universelle in Paris (1855), Goodyear was there with displays of furniture, inkstands, vases, combs, brushes and many other ordinary items, all made of rubber. Although his efforts failed to direct the flow of funds into instead of out of his coffers, the public displays did alert the world to the possibilities of this adaptable plant material.

Increasing popularity was good for Brazilian growers and traders of rubber, if not for the collectors of the raw material – the task entailed long hours and tedious, exhausting labour. Demand for rubber brought about an accelerating clearing of the Amazonian rainforests for rubber plantations. It also stimulated the search for other areas of the world where *H. brasiliensis* could thrive. And it had to be that particular rubber tree: the Brazilian Joao Martins da Silva Coutinho demonstrated its undoubted superiority.

Joseph Dalton Hooker, director of the Royal Botanical Gardens at Kew, was instrumental in spreading the cultivation of the rubber plant. His agents in Brazil produced seeds that arrived at Kew in 1876. Although only a tiny fraction germinated, seedlings were later dispatched to Singapore. They all subsequently died.

Plants in Ceylon (Sri Lanka) also struggled in the early years, although successful plantations were established eventually. The appointment of Henry N. Ridley in 1888 as superintendent of the Singapore Botanical Gardens proved to be the turning point. Ridley was a tireless enthusiast for the product: he experimented on the optimum growing conditions, on how, where and how often the plants could be tapped, and how seeds and seedlings were best transported. More efficient methods of collecting rubber were also devised, especially the use of acid to coagulate the latex. By the end of the century, Southeast Asia, including Malaysia (now Malaya), was a major source of rubber. The Chinese are a notable presence in the area mainly because of the indenture system used to meet demand. Everywhere, the creation of plantations required land clearance and a large workforce.

Use of rubber products grew continuously during the latter part of the 19th century, but the tyre, for bicycles and then automobiles, created a vast new market. In France the Michelin brothers, Édouard and André, pioneered the use of rubber for vehicles. In 1891 they patented a tyre for bicycles that could easily be removed and repaired. With the coming of automobiles, the limitations of solid tyres were shown up: they tended to break the wheels at speeds of more than 15 mph (25 kmph). Pneumatic car tyres were the answer, and in 1895 the Michelins demonstrated one. Although the tyres had to be changed every 150 km (90 miles or so), the world was convinced that this was the way forward.

Tyres, though important, are just one of a host of uses for rubber. Automobiles and most other modern machines contain rubber in hoses, fittings, washers, cable insulation and numerous other parts. It is also still used for boots and shoes, gloves, condoms and much else. All this encouraged chemists to examine the molecular basis of rubber's distinctive properties. The German chemist Hermann Staudinger showed that the substance consisted of polymers (long chains) of hydrogen and carbon, and that vulcanization, by adding sulphur, resulted in stabilizing chemical bonds. He won a Nobel Prize for his work in 1953. Although synthetic rubber is now widely used, it cannot replace the natural material for some purposes; and natural rubber is of course renewable. Brazil now produces little rubber, but the Malay Peninsula is still a major player, with China also in the game.

Banana

Musa acuminata × balbisiana

The World's Favourite Fruit

There is another tree in India, of still larger size, and even more remarkable for the size and sweetness of its fruit…. The leaf of this tree resembles, in shape, the wing of a bird, being three cubits in length and two in breadth.

Pliny, 1st century AD

Bananas are large perennial herbs; the false stem is formed from the leaf sheaths. The vast leaves have a long history of use – from roofing to wrapping to umbrellas – and some varieties are now grown in southern India for disposable 'biological plates'. For thousands of years bananas were propagated by dividing and planting the small 'pups' at the base, but successful tissue culture techniques have now been developed.

Bananas have become so ubiquitous and cheap that it is hard to appreciate that, before the 1870s, they were virtually unavailable in temperate lands and were mostly consumed at source. They originated in Southeast Asia, but the wild variety had seeds and produced little that was edible; the non-seeded, edible variety, probably a natural hybrid, has long been prized in Asia and its islands. The plant needs warmth and plenty of water (it will not grow naturally except around 30 degrees north or south of the Equator), and thrives on well-drained, rich soil. The distinctive huge leaves emerge from a large single stem. The male flowers are sterile, while the female (or hermaphroditic) flowers produce the fruits.

Because the fruit of this large herb lacks seeds, it must be propagated from lateral shoots (suckers or 'pups'), and by this means was widely spread throughout Asia and the Malay Archipelago. It reached as far as Hawaii, and Muslim traders may have introduced it into Africa (it could have been introduced even earlier via Indonesia), where it became a mainstay of the diet. Its mode of propagation means that new cultivars cannot be bred, although several natural varieties exist, the most common commercial one being Cavendish, developed in the Victorian hothouse of the Duke of Devonshire in 19th-century England. The fact that bananas are clones means they are especially susceptible to pests and diseases, a real worry in the modern world. A previous widespread commercial variety, named Gros Michel, was badly affected by a fungus in the 1930s.

Before the European age of exploration bananas had been spread throughout Asia, Oceania and Africa. The Portuguese then introduced them to the Canary Islands, and from there they were taken to the Caribbean, where their main use was as food for African slaves. The plant has the value of yielding its fruits continuously throughout the year and is one of the heaviest producers in the plant kingdom; the fruits are rich in carbohydrates, contain high levels of potassium and are a good source of vitamin C.

Bananas don't travel well, however, and needed more rapid sea transport and refrigeration to become readily available worldwide. Several American companies, including United Fruit and Del Monte, quickly exploited and dramatically

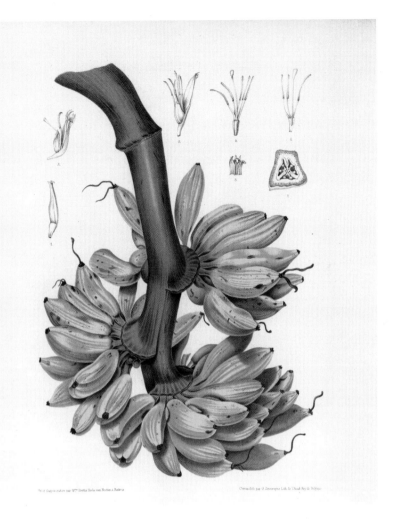

Now Capon nature, nat. Willm. Berthe Hoola van Nooten a Batavia.   Chromolith par H. Devraye per. Lith. de l'Acad. Roy. de Belgique

extended the plantations in the Caribbean islands and Central and South America. A British company, Fyffes, a partnership between an importer and a shipping magnate, did the same with the West African countries and the British market. The shipper, Sir Alfred Lewis Jones, gave away bananas to people on the Liverpool docks when his ships came in, to encourage potential customers to acquire a taste for this new fruit. These entrepreneurs established the banana as an all-year-round fruit at a time when many Europeans were seasonally limited in what they could eat.

In tropical countries bananas continue to be important local edibles, and in India the whole plant is consumed in different ways. Although mostly eaten raw in the West, bananas are often cooked in many parts of the world. In addition, plantains are also widely used in Africa, Asia and the Caribbean. They have a slightly different chromosomal arrangement, but are of the same species as the familiar yellow banana. Plantains are generally cooked and are equally nutritious.

The French Plantain (*Musa* x *paradisiaca*), here from Berthe Hoola van Nooten's *Fleurs, Fruits et Feuillages choisis … de l'île de Java* (1863), is found in Indonesia and the Pacific islands. Plantains can be eaten at various stages; unripe they are starchy, becoming sweeter as they ripen. Dried for later use, they can also be ground into flour.

Oil Palm
Elaeis guineensis
Economics Versus the Environment

Keep that wedding day complexion.
Advertisement for Palmolive Soap, 1922

The oil palm is indigenous to the tropical rainforests of West and Central Africa. A tall, long-lived tree – it can attain up to 30 m (98 ft) high and 150 years in age – it bears large clusters of fruits about the size of a plum. Before commercial exploitation in the late 19th century, the fruits were mostly gathered from wild or tended trees, and people had long discovered ways of processing the abundant oil they contained. The traditional oil, made from the fruit's flesh, is red in colour and was used locally in cooking, but did not travel well. Consequently, the first Europeans who encountered it described its properties but saw little potential value abroad. However, it was probably traded from early times within Africa, perhaps as far north as Egypt.

After harvesting and separating the seed or kernel, the fleshy part of the fruit is softened, pressed and the resulting palm oil purified. These traditional steps have been much improved by modern production methods but are still essential. Another oil is made from the kernels, and these are often exported whole for processing in the country of consumption.

As well as use in cooking, palm oils were found to be suitable for soaps, perfumes and margarine, and European demand grew. At the heart of international interest was William Lever, an entrepreneur who, with his brother James, established Lever Brothers, now part of Unilever. They began exporting palm oil from British colonies in Africa, but when securing more land for plantations proved difficult they moved the centre of operations to the Belgian Congo (now Democratic Republic of Congo, DRC). Lever combined shrewd advertising with the use of the vegetable oil in producing and selling cheap soap. Although a notable philanthropist – he built Port Sunlight as a model village for his workers in England – he also exploited cheap labour and profited from appalling working conditions in Africa.

The increasing popularity of margarine, developed in France in 1869 as a cheaper alternative to butter, led to a demand for vegetable oils. Palm oil needed less treatment than other oils, so became the product of choice. New transport methods eased its shipment and this in turn led Western entrepreneurs to look elsewhere for its commercial production. Java was the first new area, pioneered by the Dutch; it was a success, especially after the chance introduction of a new

Palmae
(Cocoineae)

Elaeis guineensis L.

The African oil palm, with its male and female flowers, fruit or drupes, and seeds. The expansion of oil palm plantations in Indonesia exemplifies the tensions between providing income for poor farmers and an export crop for a developing country, versus environmental degradation. Such concerns did not trouble the colonial planters who established oil palm as a cash crop.

variety. Soon Indonesia and Malaysia were added to the list. Malaysia is now the world's largest producer of palm oil, the downside being the destruction of much rainforest and with it the traditional home of the orang-utan. A distinct species of oil palm is found in South America (*E. oleifera*), which is employed locally but not for large-scale production. The African species is now grown there commercially in several countries.

Palm oil is low in free fatty acids (FFAs) and so is attractive to Western health-conscious consumers. Modern production is graded according to its percentage of FFAs, with the lowest reaching a premium. The oil withstands high temperatures for frying and can form the basis of 'healthy' margarines, containing vitamins A, D and E, and two essential acids.

Landscape
Plant Aesthetics on the Grand Scale

Plants make living landscapes; they colour and shape the ground on which they grow. Forests can be 'impenetrable' and the stuff of dark legend. Grasslands are 'sweeping', taking the eye to the far horizon. Plants punctuate desert landscapes and can define the boundary between land and sea. Shaping the environment en masse, landscape plants have provided an aesthetic that has been captured in art and literature. In the East, landscape painting is the most highly regarded style.

How have such plants moulded the planet? How have we adapted their landscapes for our own ends and utilized the products that these plants offer? The stately larches straddling the continents of the northern hemisphere are an important component of the great boreal forests. They have myriad uses reflecting the cultures they help shape: snowshoes for the winter months, lumber for building and exquisite Japanese bonsai. The utility of the productive timber of North America's giant redwoods vies with their impact on the emotions, exploited today through tourism. The remaining stands in California are perhaps a relic of a much wider forest, already ancient before man moved in with the axe.

Eucalyptus or gum trees, adapted to survive and even benefit from bush fires, typify the unique Australian plant heritage, and their essential oils are enjoyed. They are also now part of the new landscape of commercial tree production around the globe. Outside the plantation, the Tasmanian Blue Gum has become

an invasive weed in parts of California, crowding out native plants and creating a monoculture, reminding us how we sometimes alter and manipulate landscapes with unexpected consequences. The trees do, however, support bees and hummingbirds, and prevent soil erosion. Some rhododendrons too have gained a poor reputation. But clothing their native hillsides, with the white masses of the Himalaya as a backdrop, their colours are spectacular.

The dramatic tree-sized cacti of the American West are plants capable of life under extreme conditions. These icons of the movie industry have a longer history in the desert culture of Native Americans. They serve as a reminder of the clash of cultures as Europeans moved west across the continent. Mangroves demonstrate the peculiar plant characteristics of a life in brine. They thrive at the boundaries between land and sea, holding the two apart and preventing coastal erosion by some astoundingly beautiful adaptations. The Māori, original inhabitants of the islands of New Zealand, adopted the silver tree fern as a plant of special significance. The trunks were used for timber, while the fronds made a bed. The silver undersides, cut and laid to catch the moonlight, pointed the way – a vegetable means of communication.

Above left Rhododendron *nuttallii* – described in *Curtis's Botanical Magazine* (1859) as the 'Prince of Rhododendrons'. This small tree (10 m/33 ft) is topped by glorious, scented white flowers, the largest in the genus.

Above right The intrepid artist Marianne North captured the verdant richness of the mangrove swamp in Sarawak, Borneo (1876; detail). She traversed the country wearing rubber boots, with her skirts hitched up to her knees, and often used a boat – perhaps when she created this view from the water.

Larch

Larix spp.

Stately Conifers of the Northern Forests

We continued along the most extensive larch wood which I had ever seen,—tall and slender trees with fantastic branches.

Henry David Thoreau, 1864

The boreal forests are striking features of the landscapes of the northern hemisphere. They constitute the largest forest area in the world, and are so extensive that they have long had a significant ecological impact on climate and carbon dioxide levels. Like most forests, they contain a mixture of vegetation, but larches and other conifers dominate. They love the cool of the mountains and cannot stand wet, frozen, low-lying soil, so their reaches have waxed and waned over many millennia as temperatures and the extent of the Arctic ice have varied. Larches range widely in North America, central and northern Europe, and in Asia, from the Himalaya to Siberia and Japan. They are unusual among the conifers ('cone bearing') in being deciduous, with their leaves, 'needles', turning colour and falling in winter. Conifers are an ancient botanical group, dating back about 300 million years, and most, such as pines, are evergreen.

Larix is a relatively small genus of about a dozen species; the trees themselves are generous in size (30–50 m/98–165 ft) and fast growing. The seeds develop in the attractive crimson cones, which remain behind after the seeds are shed. The mature cones differ in each species, so are essential for botanical classification. Larch trees' fast rate of growth has made them valuable sources of fuel, and although the wood is soft, it is unusually durable – its high resin content helps make it watertight and it will not rot in the ground. This has made it a favourite for fencing, pit supports and general building purposes. It is also slow burning and deemed a good material for smelting iron. The Romans used larch to build ships, and the tradition has long continued, with Scottish trawlers and luxury yachts still built of the wood. Russians and Siberians depended on the tree to construct their houses and to warm themselves. Native Americans used it long before European settlements. Hackmatack or Tamarack are indigenous names for larches, and mean 'wood used for snowshoes'. Larch turpentine, obtained by tapping, has been advocated in both human and animal medicine.

The primary species of larch are divided into two groups on the basis of leaf markings. They are also usually familiarly known by their principal locations, for example Japanese, Siberian, European or Western (in North America). Not

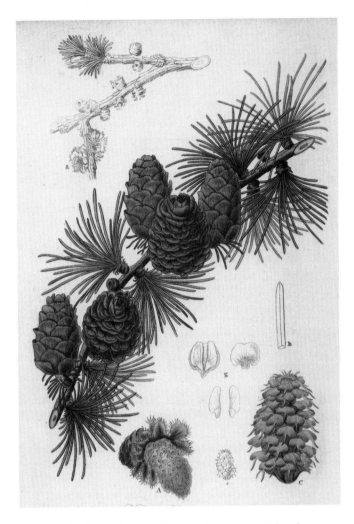

surprisingly for a plant with multiple uses, larch has been transplanted successfully. The first larch trees were probably planted in Britain only in the 17th century.
The London diarist and plantsman John Evelyn travelled to Chelmsford in Essex
where he admired a European larch (native to the Alps).

One of the most striking 'larchfests' – the spectacle of seeing the glorious
autumnal colours of larches, often among other, darker evergreens – was that
masterminded by the Dukes of Atoll on their estate at Dunkeld in the Scottish
Highlands. In the mid-18th century, John Murray, the second Duke, planted
about 150 *decidua* larches (the 'European'). His son and grandson continued the
practice, resulting in millions of trees on the estate, often on agriculturally worth-
less land. In 1895 a cross between the Japanese larch (*kaempferi*) and *decidua* was
made. This was especially vigorous and healthy, and is now known as the Dunkeld
Larch – it is the commonest commercial larch in Britain and is valuable throughout the world. At the opposite extreme, the larch is a favourite tree for bonsai.

A branch of the European larch
with cones (male, bottom left;
female, bottom right). Male
cones appear as bright green
tufts on the branches along
with the new season's needles
(detail in the top left). As with
all gymnosperms the seeds are
bare, not encased in a fruit or
vessel, although protected by
the cone.

Redwoods

Sequoia sempervirens

Titans of the Tree World

The redwoods, once seen, leave a mark or create a vision that stays with you always … they are not like any trees we know, they are ambassadors from another time.

John Steinbeck, 1962

'The coldest winter I ever spent was summer in San Francisco', said someone, even if it wasn't Mark Twain, and there's no denying the chilling damp of the coastal fog. It forms when the air above the cooler waters of the California Current comes into contact with the warmer landmass. This weather pattern perfectly suits the coastal redwood trees (*Sequoia sempervirens*). These ancient conifers are the world's tallest trees today – the record is held by the Hyperion, at 115.5 m (379 ft) high. The biggest trees, measured by volume rather than height, are the closely related giant redwoods (*Sequoiadendron giganteum*) of the western Sierra Nevada, California.

The redwoods run down the Pacific coast in a strip 8–56 km (5–35 miles) wide, from southwestern Oregon to just south of Monterey, California. They usually grow up to an altitude of about 300 m (985 ft), although some are found higher; the foliage cannot tolerate high or freezing temperatures. Redwoods and other conifers have their antecedents in the great conifer forests of the Jurassic, before the flowering plants began their successful radiations. Older, fossilized ancestors very similar to *S. sempervirens* are found over the western United States, northern Mexico and along the coasts of Europe and Asia (the other close living *Sequoia* relative, the dawn redwood, *Metasequoia glyptostroboides*, is found in a small region of Sichuan-Hubei province in China). So this coastal strip, a large but circumscribed area, is probably a relic of a once much more widespread population. Redwoods are now grown in many other areas of the world as ornamental specimens.

Only a very small proportion (estimated at 5 per cent) of the redwood's range consists of original, old-growth wood. The redwoods have been highly prized for timber, which is lightweight and resistant to decay, since the Spanish settled in the 18th century, but it was the gold rush of 1849 that began the serious inroads into the forests. Concern to preserve the majesty of the redwoods led to the founding of the Sempervirens Club in 1900 and the Save the Redwoods League in 1918. Protected areas quickly followed, but the loggers continued to cut. Tensions between the timber trade and the conservationists remain, although today the decline of coastal fogs adds a new threat. The trees' height and water requirements mean that the drier, warmer climate poses a serious problem. Redwoods were the

first trees in which the drawing in of moisture through the stomata (pores in the leaves) was appreciated. They also capture water and dissolved minerals from the fog, which then drips on to other members of their ecosystem.

Most people view the redwoods by looking upwards, but there are a few extreme botanists who have taken to the trees. They have opened up a new world in the canopy some, 90 to 105 m (295–345 ft) above the ground. Nestling in trees, some of which have been alive for 2,000 years, are more than 180 species that never touch the ground, including ferns, huckleberries, even rhododendrons, rooted in centuries-old accumulations of litter. Lichens and mosses also thrive, along with salamanders, slugs, bees and beetles. The canopy is welded together by fused secondary trunks that arise from branches, likened to the flying buttresses of the medieval churches built when these trees were already in middle age. Redwood regeneration can be vigorous – rings of trees spring from the roots of a long-dead central plant. Long may it continue.

Before the advent of steam-assisted machinery, the felling, moving and transforming of sections of redwood tree trunks into usable lumber in the sawmill was close to the limit of human capability. It could take a two-man team six days, working 12 hours a day, to cut a single tree. In Marianne North's painting *Under the Redwood Trees at Goerneville, California* (1875) the hut gives an idea of the scale of the trees.

Saguaro Cactus

Carnegiea gigantea

Icon of the Wild West

I will turn into a saguaro, so I shall last forever, and bear fruit every summer.

Myth of the Pima Native Americans

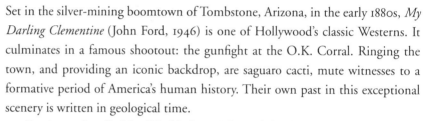

Detail from Marianne North's *Vegetation of the Arizona Desert* (1875). North's individualistic work has been criticized as satisfying neither the requirements of botanical illustration nor the high art of the later 19th century, but she has certainly captured the saguaro cactus as part of this unique landscape.

Set in the silver-mining boomtown of Tombstone, Arizona, in the early 1880s, *My Darling Clementine* (John Ford, 1946) is one of Hollywood's classic Westerns. It culminates in a famous shootout: the gunfight at the O.K. Corral. Ringing the town, and providing an iconic backdrop, are saguaro cacti, mute witnesses to a formative period of America's human history. Their own past in this exceptional scenery is written in geological time.

Cacti are primarily New World plants. They exhibit the classic suite of adaptations – the 'succulent syndrome' – of arid landscapes. The saguaros have immense stems, which can photosynthesize and store huge quantities of water from the brief summer rains. They use a special type of photosynthesis (Crassulacean acid metabolism), which sees the uptake of carbon dioxide at night, thus minimizing water loss through the stomata during the day. Anchored by a taproot, they also have a network of smaller roots near the surface ready to absorb any water when it does rain. A thick, waxy cuticle covering the plant, and leaves reduced to whorls of spines also help to cut down water loss. The spines deter larger herbivores and provide shade for the stem.

The tall (over 15 m/50 ft) and columnar saguaro cacti are stately, unhurried plants, living for 175–200 years. Slow to reach flowering maturity (30–35 years and at 2 m/6½ ft high), only at 50-plus years old do the candelabra-like branches begin to appear. The plants produce flowers on the stem tips – more stems means more flowers and increased reproductive success. From April to June, the white, waxy, funnel-shaped blooms open at night, releasing a ripe-melon fragrance to attract their nocturnal pollinators, the lesser long-nosed bats. White-winged doves and insects take over the next morning, before the flower fades. Each blossom lasts only 24 hours. The red fruits with their numerous fat-rich seeds follow from May to July. They are a useful food source for a range of animals including the Gila woodpeckers, which make their home in the stems, pecking in and down to form a pocket. Once these are vacated by the woodpeckers, elf owls move in. In response to an injury, saguaros form a hard callous around the wound. When the plant dies, these nesting boxes remain and are known as 'saguaro boots'.

Arborescent (tree-like) saguaro cacti are part of a unique flora concentrated in the Sonoran Desert – they live nowhere else but here and in places form groves. It is thought that the central Mexican ancestors of the saguaro responded to and evolved with the desertification of the region, 15 to 8 million years ago. They are an intimate part of the biome, connected with other plants such as the palo verde 'nurse trees' that protect saguaro saplings until they outgrow and outlive them.

For much longer than today's tourists and yesterday's gunslingers, the Native Americans in the Sonora region have celebrated and utilized the saguaro. Using a woody rib from a dead plant, the Pima and Tohono O'odham peoples harvest the fruit, eat it fresh and save the seeds for grinding into meal. They also make a syrup, which is fermented into wine to be drunk at rain ceremonies; an age-old celebration of the renewal of life in the desert.

Saguaro cactus from *Curtis's Botanical Magazine* (1892). Joseph Dalton Hooker reported proudly of the saguaro in the Kew Palm House that 'the flowering of this wonderful plant in England must be considered one of the triumphs of Horticulture'. Cacti took their place among the exotics in botanic gardens and collections of wealthy amateur growers.

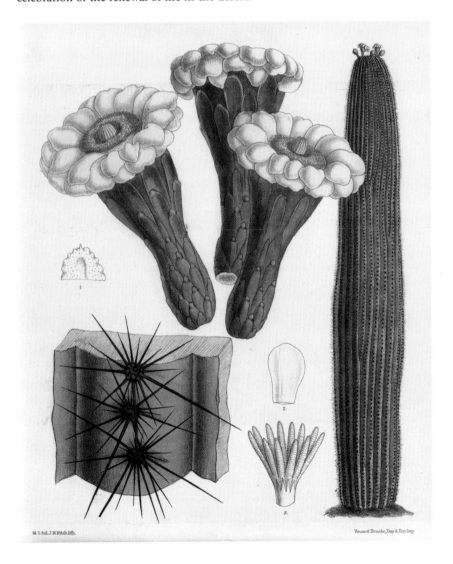

M.S.del. J.N.Fitch lith.   Vincent Brooks, Day & Son, Imp.

Silver Tree Fern

Cyathea dealbata

Shining Māori Symbol

Ponga ra! Kapa O Pango, aue hi!
[Silver fern! All Blacks!]

All Blacks Rugby Team, New Zealand, pre-match haka, 2005

New Zealand separated from Australia about 80 to 100 million years ago, giving the islands plenty of time to develop a special flora and fauna. They were so unusual in fact that when Sir Joseph Banks visited in 1769 with James Cook on the *Endeavour*, he recognized only 15 of the first 400 plants he examined – 89 per cent of its native flora is exclusive to New Zealand. The animals were once equally exotic, some of them already extinct before the Europeans arrived.

Ferns, which with related plants constituted more than 12 per cent of New Zealand's flora, are very old: evidence of their ancestors dates from the middle Devonian period (around 385 million years ago) and tree ferns are present in the mid-Triassic (around 235 million years ago). Today ferns exhibit a vast variety of form and habitat, but they have changed so little through evolutionary time that one of their primary biological uses is to study the basic mechanisms of life in early forms. Although 'trees' in size and general appearance, tree ferns rarely branch and do not have true bark, instead being protected by the roughened bases of old fronds, often with an additional covering of lichens, moss or other organisms. Their root system is very compact, which means that larger examples (up to 25 m/80 ft) generally must be supported by adjacent plants. Like other ferns they reproduce by spores formed on the back of their fronds. The two principal genera of tree ferns are *Dicksonia* and *Cyathea*, both highly successful and widespread in their geographical distributions.

The silver tree fern, *C. dealbata,* is one of a number in New Zealand, but was especially prized by the Māoris, who permanently settled the islands from East Polynesia in around AD 1250–1300. It can grow as high as 10 m (33 ft), with fronds 2–4 m (6½–13 ft) long radiating from the top. Known as kaponga or ponga to the Māoris, the silver tree fern takes its English name from the whitish-silver underside of its fronds, which can give the plant a majestic appearance in moonlight. Traditionally, this characteristic was exploited by laying the fronds upside down on the ground to mark trails in the dark. The Māoris also used the plant for building rat-proof food storage houses and making utensils. The tough, durable trunks have since been employed in fencing and landscaping and for making vases and boxes. The pith and young fronds can been cooked (the Māoris did so), and medicinal recommendations include a wound dressing for easing boils or cuts and a treatment for diarrhoea.

The silver tree fern was part of the fern craze – pteridomania – that swept Victorian Britain and to some extent North America. Specialist dealers arose and their wares fetched high prices as wealthy people in the northern hemisphere hoped to meet the challenge of growing exotics out of their natural environments. The silver tree fern occurs unevenly throughout New Zealand's islands, and all tree ferns need to be protected against export for gardeners and habitat loss for a variety of reasons. In New Zealand, *C. dealbata* has become a national icon, its distinctive silver leaves gracing the black jerseys of the national rugby union team, the All Blacks.

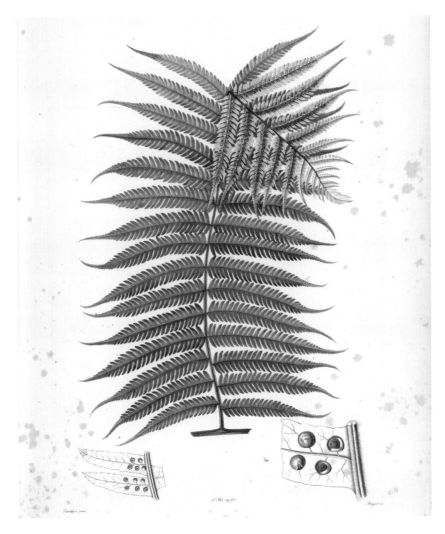

Frond (with spore details) of New Zealand's silver tree fern, from the *Voyage de la corvette L'astrolabe exécuté pendant les années 1826–1829* (1833). This voyage was the first of two French expeditions to the southern hemisphere under Captain Dumont D'Urville which were immensely important for bringing back new material to European collections.

Eucalyptus

Eucalyptus spp.

Australia's Signature Tree

Eucalypts are Australia's major source of hardwood. Not surprisingly in the colonial period, when many viewed the tropics as virgin territory for plantations of all kinds, these trees were trialled in botanic gardens (this photograph was taken in Bogor, Java). Like all non-indigenous species, eucalypts can become serious invasive nuisances.

Kookaburra sits in the old gum tree.

Marion Sinclair, 1932

Eucalyptus is a large genus of trees, with about 500 species that come in a great variety of sizes and shapes. Native to Australia and a few Pacific islands, they are now to be found worldwide, so adaptable are they. A number of common names such as gum trees (of several types), peppermint trees and bloodwood trees are indicative of their many characteristics and uses.

Eucalyptus trees evolved several million years ago, when Australia was already separated from the larger Asian landmass but was much wetter than it has since become. As conditions in Australia changed, several members of this resilient genus developed adaptations that have permitted it to become so widespread and dominant. First, very deep and extensive root systems allow trees to survive where surface water is scarce. Second, it can thrive after forest fires, often caused by lightning. Firing actually helps the trees' seeds to germinate and the ash-rich forest floor offers the perfect conditions for a new crop of eucalyptus saplings. In addition, buds just below the bark of a mature tree are stimulated by fire, so the eucalyptus can rise Phoenix-like from the ashes. So many species of the genus adapted that there was hardly an ecological niche in Australia that it could not colonize. One species, *E. regnans*, is the tallest flowering tree in the world, and the eucalyptus regularly towers over its competitors in forests.

The Aborigines exploited the eucalyptus, making from it spears, boomerangs and rough canoes. After Europeans settled in Australia from the late 18th century, the tree's vigour and range of applications were recognized and it was exported to many other parts of the world. There are specimen trees virtually everywhere and commercial plantations in Brazil, the United States, northern Africa and India. Many species are extremely fast growing – saplings of some can reach 1.5 m (5 ft) in the first year, and 10 m (33 ft) by their tenth – and make excellent firewood. Indeed, the biomass obtained from an acre of eucalyptus rivals that of any other tree. By planting it in marshes its efficient water uptake has been used to drain wet places, and its volatile gums are disliked by mosquitoes, further helping malaria control.

Eucalyptus yields a number of valuable products. Its oils are a distinctive feature of the trees' adaptive mechanisms and some, such as eucalypol (cineole),

Nº 501.

Eucalyptus persicifolia.

G.C.Fecit.

are used medicinally and occasionally in perfumes. The Blue Mountains near Sydney take their name from the pungent haze created by the extensive eucalyptus stands there. The leaves, bark and fruit, for example in *E. gunnii,* yield a variety of pigments that are used in dyeing, and the genus is an important source of tannin. The trees make good windbreaks while the hard woods can be employed in making musical instruments and they are a central core of chipboard; the pulp is a major ingredient in paper manufacture.

Although an extremely important commercial tree, eucalyptus's aggressive vigour and spread around the globe give it a bad name among conservationists. Its success is aided by the fact that its oils make it relatively impervious to predators. One animal that can digest its leaves is the koala, a marsupial that evolved only in Australia, and the sight of these sedentary eating machines feeding on eucalyptus leaves is a famous icon of Australia's unique ecology.

The flowers of *Eucalyptus persicifolia* from the *Botanical Cabinet* of the Loddiges nursery in Hackney, London. Part botanical periodical, part subtle advertisement, the fine plates were engraved by George Cooke. This ornamental eucalypt was grown in the nursery's conservatory to protect it from frost; its rarity may have commended it to the discerning grower.

Rhododendron

Rhododendron spp.

Flowering Mountains

… three Rhododendrons, one scarlet, one white with superb foliage,
and one, the most lovely thing you can imagine.

Joseph Dalton Hooker, 1849

A taxonomic nightmare, the *Rhododendron* genus is one of the largest in the entire plant kingdom: it contains more than 800 species with probably still others yet to be described. Largely a northern hemisphere plant, although Australia has its own species, rhododendrons vary enormously in size, shape and habitat. Some are tiny Alpine-like plants, others substantial trees up to 30 m (around 100 ft) tall, such as the aptly named *R. giganteum.* They can form dense, luscious tropical landscapes and are found in the high Himalaya at the extreme limit of vegetation. They can even be epiphytes, growing on other trees. Their root systems are usually compact, although in drier climates some species have more spreading roots. Almost all share one characteristic: they need acid soil.

With many species exhibiting striking, waxy, deep green leaves and spectacular, often fragrant blooms in a broad spectrum of colours from light pastels to deep crimsons and even blues, they have become plants for gardens as well as landscapes. A small eastern European species was cultivated in the early 17th century, and the French naturalist and traveller Joseph de Tournefort admired *R. ponticum* in Anatolia in the early 1700s. European fascination began in earnest with the discovery of North American species that were brought back to Europe throughout the 18th century. When Linnaeus classified rhododendrons he placed azaleas in a different genus, but the two are now grouped together, with azaleas occupying several subgenera of the larger grouping. Equally prized in modern gardens, azaleas are generally deciduous or partly deciduous and are smaller and less aggressive than their cousins.

When Joseph Dalton Hooker was botanizing in India, including the Himalaya, in the late 1840s, he was astounded by the rhododendron landscapes. He identified 28 new species in Sikkim alone, describing them in his magnificent book, *The Rhododendrons of Sikkim-Himalaya.* It was Hooker who systematically investigated the geographical and morphological range of this giant genus, and he continued to study them after his return to England, where he succeeded his father William Hooker as director of Kew Gardens. Between them, they introduced a few Asian species to Britain, and then to Europe and North America,

---

although most Asian species have been introduced in the past century, many from western China.

Since rhododendrons hybridize easily, new varieties have been regularly developed, as new characteristics were sought, including greater hardiness, flower colour or size. Often it has been a case of East meeting West. For instance, an early British hybridization, from 1814, crossed a Turkish variety (*R. ponticum*) with an American species (*R. periclymenoides*).

Rhododendrons grow well where conditions are roughly similar to the soil and temperature of their native habitats. Like many plants bred for gardens, they suffer from a variety of pests and diseases, but this did not deter fascination among gardeners. Those without the necessary acid soil often imported peat, depleting ancient peat bogs. At the same time, several species, particularly *R. ponticum*, formerly used as the basic grafting stock, have naturalized easily and spread rampantly, creating new, unlooked-for landscapes. In Scotland pigs are employed to forage in unwanted rhododendron stands to weaken the root systems and help keep the plants under control. Although largely ornamental, rhododendrons (like virtually every other plant) have been used medicinally, especially in Asia, and sometimes the young leaves are cooked.

Above left 'Snow beds at 13,000 feet, in the Th'lonok Valley: with Rhododendrons in blossom, Kinchin-junga in a distance': the frontispiece to Joseph Dalton Hooker's *Himalayan Journals*, vol. 2 (1854). Hooker was captivated by the scenery and plants, but noted the scent could be 'too heavy by far to be agreeable'.

Above right 'Rhododendron dalhousiae in its native locality' served as the frontispiece to Hooker's *The Rhododendrons of Sikkim-Himalaya* (1849–51). Hooker's foray into the eastern Himalaya yielded 25 new rhododendron species. For the book, Hooker's original artwork was transformed into superb lithographs by Walter Hood Fitch.

Mangroves

Rhizophora and other species

Between Land and Sea

*The red mangrove grows commonly by the seaside, or by rivers or creeks. The body …
always grows out of many roots about the bigness of a man's leg…. Where this sort of
tree grows it is impossible to march by reason of these stakes.*

William Dampier, 1697

Swathes of mangrove forests cover some 15,000,000 ha (37 million acres) around
shorelines in the tropics and subtropics. A diverse group of specialized plants,
around 70 species and hybrids of trees (up to 30 m or 100 ft tall), shrubs, ferns and
a single palm are considered 'true' mangroves, of which 38 are 'core' species domi-
nating the forests. Mangroves form shoreline ecosystems and support a wealth of
other plants and animals.

Their presence in the tidal littorals protects the edge of the land from the
perpetual wearing of the sea, the ravages of storms and the effects of extraordinary
events. On 26 December 2004 massive waves of immense height and speed were
generated in the Indian Ocean, caused by a huge undersea earthquake. Nothing
could entirely absorb the force of the waves where they reached shores, but damage
was mitigated in inland areas behind intact mangrove forests. Where mangroves
had been cleared, often to provide a reliable source of income for local people
through shrimp farming, the devastation was extreme.

From the air the intertidal home of the mangrove forest is clearly demarcated:
a verdant splash of green between the sea and the interior. But these plants are
probably best seen, like seaports, from a boat. From the water it is possible to
appreciate the adaptations that have allowed these plants to reside in this often
muddy and unstable transitional zone between land and salt water, and the different morphologies of the mangrove forests.

Fringing mangroves live along narrow strips that follow the shoreline, lagoon
or numerous channels cut by river deltas. Although they can cope with seawater,
mangroves thrive where salinity is diluted by freshwater from rivers or high rainfall. In shallow basins, broader forests develop, sheltered from daily tidal ingress.
Patchy 'overwash' mangroves inhabit islands or headlands that are inundated at
high tide, making it difficult for leaf litter and other useful debris to build up.

Mangroves deal with the conditions in their brackish world in various ways.
All reduce the amount of salt absorbed through the roots as much as possible;
Rhizophora are keen excluders. Others, including *Avicennia*, excrete it through
salt-secreting glands that leave characteristic crystals on the leaves. A third group

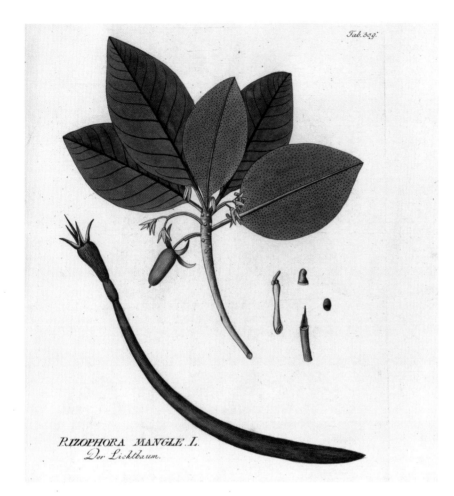

Tab. 329.

RIZOPHORA MANGLE. L.
Der Lichtbaum.

The leaves and propagule of the red mangrove (*Rhizophora mangle*). Red mangroves are indigenous to the western coast of Africa, and the eastern and western coasts of subtropical and tropical America, between 28 degrees north and south of the Equator. Their ability to trap sediments helps filter the water, and by building up deposits into peaty soils they act as a carbon sink.

of plants allows the salt to build up in the leaves and then sheds them as necessary. Regeneration of mangroves is essential for the health of the forest, but the shifting bottom and movement of water make treacherous settings for seedling growth. To get around this, the plants have developed various means of allowing seeds to germinate while still attached to the parent. The resulting propagules can become quite large, those of *R. mucronata* are a metre (3 ft) in length.

Mature plants must also cope with soft, waterlogged ground. *Rhizophora* employ stilt roots from high up the stem, like guy ropes; additional roots can descend from branches. *Heritiera* develop massive buttress roots over a large area to anchor the plants along typhoon-prone shorelines. Held fast, such large living structures must be able to breathe in the mud. Some have lenticels – the equivalent of leaf pores – on their stems. Breathing roots of various shapes and sizes – knee roots, peg roots – stick up from hollow horizontal roots on others. The fascinations of this landscape may come to their rescue, as the aesthetic as well as the practical potential of these plants is increasingly appreciated.

Revered and Adored
From the Sacred to the Exquisite

Plants meet the needs of our senses and spirit as well as our bodies. The veneration of plants is evident in some of the earliest artifacts, and the mythology of ancient societies is replete with plant deities. The texts of today's major religions refer to the importance and sanctity of plants. Nature worship, tree hugging, plant appreciation societies and flower festivals – all flourish as we continue to esteem plants for their varied meanings and associations, and the pleasures they bring.

The lotus is hallowed for its ability to grow in stagnant water and rise seemingly from the dead. It is a potent symbol of rebirth and purity; its iconic buds and opened flowers ornament the Hindu and Buddhist traditions. The rose is the quintessential flower of love. It is a mutable emblem that was favoured wherever it naturally grew. As different species were brought together from East and West, its very form changed over time. The pursuit of scented, repeat-flowering roses of myriad hues yielded a rose-fancier's paradise.

The tipping point between a healthy passion and floral madness has happened more than once and in disparate parts of the world. Gorgeous peonies – the Chinese flower of riches and honour – sold for enormous sums in Tang dynasty China as gardeners vied to produce huge flowers. Risky plant-hunting expeditions were required to satisfy demand in the West. Orchids suffered at the hands of rapacious connoisseurs who sought new plants for their collections. On a more

TULIPA GESNERIANA DRACONTIA
TULIPA GESNERIANA VARIEGATA

Tulipani mostruosi
Tulipani brizzolati

scientific level they also drew in those who wished to understand the often unique relationships of plant and pollinator. 'Tulip fever' raged during the Dutch Golden Age and led to an unsustainable futures market. The Dutch were consumed by the same passion as the Ottoman Turks, who had scoured and stripped the mountainsides of Central Asia for the smaller, delicate species tulips. Tulips inspired many artists, as did Chinese plum blossom, a harbinger of spring even as winter seems still to grip the landscape. Dainty flowers open before the leaves on bare stems – the delicate brush stokes needed to capture their fragile beauty honed the skills of many a fine painter and calligrapher.

And what was the forbidden fruit of Eden? Apples and pomegranates have both shared the epithet, and have links with fertility, immortality and abundance. Date palms too are symbols of everlasting life. These desert icons have long associations with the concept of the 'tree of life', their fruits succouring and sweetening and their leaves a potent symbol of peace. Sanctity and ritual purity are the attributes of frankincense, the dried gum of *Boswellia* trees. Burnt during religious rites for millennia, its scent is often essential to the shared experience of worship.

Above left The showy flowers (a double-flowered variety on the right) and developing fruits of the pomegranate, with the ripe fruit below, split to show the juicy arils in their compartments.

Above right Two choice tulips, 'Dracontia' (bending to the right) and 'Variegata', from the Italian physician and botanist Antonio Targioni Tozzetti's *Raccolta di fiori frutti ed agrumi* (1822–30).

Opposite Orchis morio (or *Anacamptis morio*) – the green-winged orchid – one of the native British orchids Charles Darwin used in his experiments to determine the role of insects in pollinating these enchanting flowers.

Lotus

Painted in India between 1860 & 1870
by Mrs Fanny C. Russell   21 June, 1928.

Lotus

Nelumbo nucifera

Sacred Flower of Purity and Rebirth

*When we can imagine ourselves free from guilt, we bloom as the lotus …
in the summer dawn.*

D. T. Suzuki, 1957

In the light and warmth of the new day the lotus bud unfurls its petals accompanied by the most discreet susurration. Through the day, the lotus's delicious scent intensifies, attracting the pollinating insects that will spend a night within the closed flower. They are kept on the move by the warmth the plant generates. After three days the glory of this most bewitching of blooms begins to fade. The petals fall, leaving an iconic conical boss. This will swell and grow, tip on its side and, once mature, fall into the water, its contents having developed into bean-like seeds.

If the petals are ephemeral, the seeds have an astonishing longevity. Seeds dated to over a thousand years old have been successfully germinated. And the flower's delicate beauty atop its stalk contrasts with the rhizome below, for this emblem of purity thrives in stagnant water. The circular leaves, supported on spiny stalks above the water line, have water-repelling adaptations on the upper side, which shrug off much of the rainwater that falls. What is caught in the dished centre is absorbed through a pore connected to special channels in the stem that continue through the rhizomes. This system provides additional water and a means of gaseous exchange. It is testimony to the lotus plant's size that although water-dwelling, it needs yet more water. Capable of rapid growth, the lotus can quickly take over the shallows and marshes of rivers and lakes, producing massive clones.

All this has continued for millennia. What we see today as wild lotuses dotted in the warm temperate and tropical parts of Asia – Iran to Japan; Kashmir and Tibet through New Guinea to northeastern Australia – are relics of a once more widespread population. Over the span of geological time, following a rise to prominence in the Cretaceous, the lotus lost out in the desiccation of a colder, drier earth. Our love for the plant may have helped reinstate it.

In many of its native regions the lotus became indelibly associated with ancient cultures and their religions. Whenever they travelled, in conquest or to spread the word, people carried their lotuses, muddying the patterns of natural and artificial dispersal. Everywhere, this was a plant of rebirth. Its renewed growth in dry ponds after the rains endowed the lotus with a magical quality. The lotus has long featured in the life of Mesopotamia. It was associated with the cult of Inanna, fertility

Opposite 'Lotus' – a watercolour painted in India between 1860 and 1870. Appreciation of the lotus extends across Asia and festivals celebrating the opening of the flowers have become important tourist attractions. Lotus ponds are part of the West Lake Cultural Landscape of Hangzhou, China, in an area of the Yangtze River delta that has been celebrated since the Tang dynasty and is now a UNESCO World Heritage Site.

Below The idol 'Pussa' depicted sitting on her lotus in Athanasius Kircher's *China Illustrata* (1667). Trying to make sense of the reports of Chinese Buddhist images, Kircher equated them with the Egyptian and Greek deities and so the Pussa becomes the Chinese equivalent of Isis or Cybele.

Opposite 'Tamara of India' from Richard Duppa's *Illustrations of the Lotus of the Ancients and Tamara of India* (1816). The seeds in their central boss are clearly shown amidst the rosy petals. Duppa, trained in law, wrote on botany, art and politics, and spent time painting in Italy, combining his various skills in this engaging volume.

Below A late 19th-century Indian rosary made from the seeds of *Nelumbo nucifera*. The followers of all the major religions in India use rosaries, and they are often made of seeds, the sacred nature of the lotus perhaps making this an auspicious choice.

goddess of the city of Uruk, and reproduced in jewelry and on seals. The lotus-shaped sceptre was a symbol of ascendancy, maintaining its prominence through the region's successive empires until Alexander the Great vanquished the Persian king Darius III in the late 4th century BC. Under Persian influence the lotus seems to have joined the existing blue and white waterlilies in the waters of the Nile in the first half of the 1st millennium BC, supplanting them in the rites of Isis. As the goddess of birth and renewal, Isis became a leading Egyptian deity in the receptive Greco-Roman pantheon.

The lotus would also become crucially important to the beliefs of the Indian subcontinent. As recorded in the Vedas that underpin Hindu mythology, Vishnu and Lakshmi are central to the creation story. Afloat in the void on a giant snake Vishnu sprouts a lotus from his navel. The flower opens to reveal Brahma, who then sets to work to make the universe. Goddess of fertility and prosperity, Lakshmi is born holding, or like Brahma is seated on, a lotus. It becomes her flower. The lotus is no mere adornment. As the vedic canon developed, the 'lotus of the heart' became the container of one's inner cosmos, the destination of the spiritual quest of life.

The lotus also bore the Buddha, just as it held the earth above the maelstrom of the universe. Buddhism in its various teachings spread from India to China, Korea and Japan, Tibet and Sri Lanka, adapting and taking on local associations, but the lotus remained a central image. It was cherished and cultivated for its essential symbolic role in the achievement of enlightenment. By seeking to rise as the lotus bud does from the mud (the evil of human ways) and to float above all worldly things, the follower of Buddhism could approach nirvana. The *Lotus Sutra* (*c.* 1st century BC) was one of the most important texts of Mahayana Buddhism, which espoused a simplified doctrine. In the 13th century AD, the Japanese Nichiren founded a school that further elevated this sutra.

After the Meiji restoration (1868) and the opening of Japan to foreigners Japonisme became increasingly popular in Europe and America. The lotus was an integral part of this new fascination with the East, and was incorporated into Art Nouveau's later repertoire of nature-inspired glass, ceramics, jewelry, furniture and fabrics. It is now grown wherever possible as an inspiring ornamental. Replete with spirituality, the lotus also nourishes the body. The rhizomes and seeds (and to a lesser extent leaves and flowers) are eaten throughout eastern and southern Asia, and the West is slowly waking up to the taste of these ancient beauties.

Cyamus Nelumbo. Dʳ SMITH.
XIII
7

TAMARA

of India.

Date Palm
Phoenix dactylifera
Bread of the Desert

Shake the trunk of the palm tree towards thee: it will drop fresh, ripe dates upon thee. Eat, then, and drink, and let thine eye be gladdened!

Qur'an 19:25–26

Engelbert Kaempfer travelled extensively in Russia, Persia and Asia in the late 17th century and wrote about his adventures in *Amoenitatum Exoticarum* (1712). He witnessed the caravans that arrived in Isfahan, Persia's capital – the merchants had long enjoyed the nutritious combination of camel milk and dates.

The shimmering mirage of an oasis fringed with date palms has tricked many a thirsty desert traveller. The real tree succours and shades, and is found in rocky outcrops as well as the shifting sands of Southwest Asia and North Africa. The date is the lynch pin of the unique oasis ecology – the palms are nurtured by the groundwater that lies within reach of their roots and tolerant of some salinity.

With its tall, slender trunk up to 30 m (almost 100 ft) high, topped with a crown of fronds, the date palm also offers a protective canopy in irrigated plots. Beneath the palm there is room for fruits, cereals and vegetables, and other useful crops. Such 'date palm garden' agriculture, found from the early Bronze Age, around the early 3rd millennium BC in Mesopotamia, added to the date's already considerable bounty. Sap from the stem can be drunk fresh or fermented to make date wine; palm pith makes flour and palm hearts serve as vegetables. The super sweet fruits, held high up in great bunches, are a cardinal foodstuff. When ripe and dry (varieties are classified by their softness and dryness) they are easy to preserve, wonderfully transportable and highly nutritious. When pressed, the fermented dates produce a syrup, the honey of the biblical 'milk and honey'. The trunks, leaves, kernels and their oils serve as timber, roofing, fuels, multipurpose fibres – especially for basketry – fodder, caffeine-free 'coffee' and soap. Little wonder that the date palm was raised to divine status and has been identified with the trees of life, abundance and riches.

The Sumerians celebrated dates, using the iconic trees on their cylinder seals. The Egyptians created columns topped with palm capitals and the god of eternity, Heh, grasped notched palm branches used to record time. The date palm became a leading motif of the ancient world even in places where the climate prevented cultivation. In highly stylized form it featured in reliefs in the stupendous state apartments of the Northwest Palace at Nimrud built for the Assyrian king Ashurnasirpal II in the late 9th century BC.

The palm's ancient association with everlasting life may relate to its ability to cope with fire damage (hence *Phoenix*), and to the regular growth of new fronds throughout the year. What would not have been so apparent is the special nature of the cells of the stem. These are not immortal but essentially last throughout the roughly 150-year lifespan of the tree. The date palm is best propagated from off-shoots, which sprout at the base of the parent plant. These clones allow selection for female over male trees, since only the females produce fruit.

A frond, carried in the hand of the victor, was a mark of triumph in Greek and Roman athletic and other competitions. The date palm was also a significant symbol in all the monotheistic faiths of the Holy Land. Its fronds are an important element in the Jewish festival of Sukkot. Those who celebrated Jesus' entry into Jerusalem greeted him with waving palm fronds, and early Christian martyrs are often depicted with one. According to Muslim tradition the date was made of the dust left over after Adam's creation, and it spread with Islam to Spain. Among other instructions, Islamic armies were ordered not to destroy existing palm trees. Dates were an intrinsic part of the Arabian identity – traditional buildings on the Arabian Peninsula are still made from the leaves of date palms. And what could be more natural to break the month-long fast of Ramadan, each day at sunset, than these marvellous morsels of the desert?

Frond, flower and fruit of the date. The tradition of planting date palms in palace gardens in Persia is reflected in the fabulous garden carpets that were woven to provide eternal spring within.

PHOENIX DACTYLIFERA

Frankincense

Boswellia sacra

The Odour of Sanctity

Trees have their allotted climes … to the Sabaeans alone belongs the frankincense bough.
Virgil, 1st century BC

Resins are both valuable and useful – think of amber, tar and pitch, turpentine and rosin. Many come from conifers, and some were appreciated for their scent. Other trees, too, were tapped for their fragrant resin, to be burnt as redolent incense, or used in medicines and as perfume. Among the most prized was frankincense (or olibanum), taken from the four species of *Boswellia*. Of these, *Boswellia sacra* was the most prestigious. A marvellous mystique enveloped the land of the Sabaeans (Yemen, Oman) where *B. sacra* grew – the air was said by the Greek Agatharchides (2nd century BC) to be permeated by the sweetest scents.

Gold, frankincense and myrrh (resin from *Commiphora* trees) are hallowed as the gifts presented to Jesus by the Magi. True or not, they represented three of the most valuable items of the time. By the mid-4th millennium BC frankincense was being imported into Mesopotamia, possibly from the southern Arabian Peninsula. The overland Incense Route may have stimulated domestication of the single-humped camel as beast of burden, and certainly by the late 2nd millennium BC desert communities were controlling the lucrative trade in aromatics and spices that crossed the sands to markets in Egypt, the Levant and beyond, ultimately as far as India and China. Towns en route enjoyed a healthy income from the taxes, and the Nabataeans grew rich on the trade, building their city at Petra. Frankincense was sufficiently valuable that those who sourced and traded the best resin wanted to keep the exact location of their trees to themselves. Historical authors confused matters, some thinking that juniper provided the pure incense of frankincense.

The flame, smoke and scent of incense were all significant in the rituals, sacrifices and oblations of many ancient cultures. Egyptians and Greeks shared a similar belief that the scent protected against evil spirits and signified the presence of the gods, who were by definition fragrant beings. Incense was itself an offering: instructions for its composition were set down in the Hebrew Bible. It also formed part of the preparatory stage of Roman animal sacrifice. The early Christian church abjured incense, but it was used again by the 5th century. Burnt in thuribles it was wafted to cense the priests, scriptures, altars and the elements of the Eucharist.

A censor, wafting incense, stands next to a hookah, as recorded by Engelbert Kaempfer. In Christian rites, incense was used more in the Eastern Church than the Western, and Protestants dispensed with it after the Reformations of the 16th century.

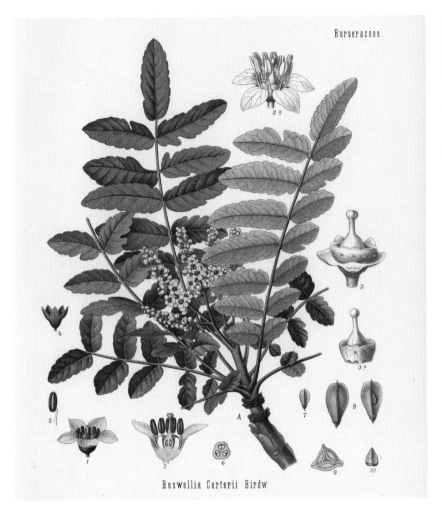

Burseraceae.

Boswellia Carterii Birdw.

At the height of the frankincense trade, Yemen and Oman were renowned for their perfumed air. *Boswellia* was harvested and burnt there, and it became a major trading region for spices and perfumes from the neighbouring African coast, and from India once the pattern of the monsoon winds was appreciated.

Today the tree grows in a limited ecological niche in the 'frankincense region' of the southern Arabian Peninsula. Trees lodged in clefts in the limestone escarpment, inland from the coast, enjoy the best water supply in this otherwise arid region. Their multiple trunks are covered with papery, peeling bark. Slashed, it releases the oily gum-resin, and if the cutting is moderated the tree will recover and remain productive. The best quality resin is allowed to run down the tree as tears. It dries and hardens into highly aromatic whitish pearls or beads. These can be softened and worked into larger quantities or ground to powder for blending. Incense is often made up of frankincense, other aromatic resins and spices.

It seems likely that increasing dryness, intense exploitation at the height of the trade in the 1st and 2nd centuries AD and then a falling price thereafter may have caused a decline in the number of *Boswellia sacra* trees, and today numbers are declining. The trees of Dhofar in Oman are now part of a UNESCO World Heritage Site: the Land of Frankincense.

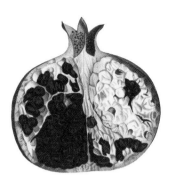

Pomegranate
Punica granatum

Fertility, Abundance, Renewal

And they made upon the hems of the robe pomegranates of blue, and purple, and scarlet, and twined linen.

Exodus 39:24

Long used medicinally, today the pomegranate has achieved a new popular status as one of the 'superfoods', said to be effective against heart disease.

Break open a pomegranate's leathery rind and inside are the tightly packed compartments of a jewel casket. Each small seed is surrounded by a beautiful, juicy aril, an extra seed coat which in the pomegranate forms the pulpy, enticing flesh. The colour of the flesh, depending on variety, ranges from deep red to the merest hint of translucent pink. Like colour, the pomegranate's taste also depends on variety. Since its domestication from the wild orchards of Transcaucasia, north-eastern Turkey and the southern Caspian region, the balance of sweet and sour has been selected for and perpetuated by taking cuttings from favoured parent trees. These pretty trees, with their shiny deep green leaves and long-lasting, extravagant red flowers, also pleased their cultivators.

Wherever it was grown, the pomegranate entered the psyche as well as the garden. The abundance of seeds in each fruit often symbolized fertility, regeneration and abundance. The Zoroastrians of the central Iranian plateau used pomegranates in initiation and marriage ceremonies. The fruit appears frequently in the Hebrew Bible, often grouped with the olive and vine. There are descriptions of its form embroidered on priestly and regal garments, and decorating Solomon's Temple. The shape was replicated in oil lamps and embossed on coins. Said to contain 613 seeds, each corresponding to the Torah's 613 *mitzvot* (commandments), pomegranates feature in the food celebrations of Rosh Hashanah.

The Phoenicians may well have taken the pomegranate to North Africa and the western Mediterranean. In Greek myth, pomegranates are chiefly associated with Demeter, goddess of agriculture and the harvest, and her daughter Persephone. Hades, who had abducted Persephone, tricked her into eating some pomegranate seeds. Since eating anything in the underworld condemned the consumer to remain there, she was required to stay in the kingdom of the dead for part of the year. Her return to the earth each spring signalled new life of the forest and field. The fertility symbolism also permeated Greek understanding of the female body. The pomegranate's blood-like, many-chambered interior was likened to the womb. Hippocratic recipes to assist conception and for fever after childbirth included pomegranate juice.

Traditionally a pomegranate tree is said to have arrived in China's Han court from Kabul with the returning envoy Zhang Qian, around 135 BC, although a document from a tomb sealed in 168 BC also names the plant. It was greatly appreciated for its flowers and commemorated in poetry as the essence of red beauty during the Six Dynasties (AD 220–589) – the flowers' shape and colour were likened to the skirt of a courtesan's dancing costume.

The Christian Church assimilated the pomegranate's association with resurrection, and its red juice symbolized Christ's shed blood. In religious art the Madonna often holds an infant Jesus clutching the fruit (for instance *Madonna of the Pomegranate*, by Botticelli, *c.* 1487). This medieval trope was stitched into a series of famous tapestries, *The Hunt of the Unicorn* (1495–1505). In the final panel (thought by some to stand alone) the captive unicorn is tethered to a pomegranate tree. This can be read both as Christ on the cross and as a celebration of matrimony, with the pomegranate as the emblem of fertility and the indissolubility of the union. In 1509 when she married Henry VIII of England, Catherine of Aragon chose the pomegranate for her coat of arms. It was to no avail. Henry had their marriage annulled in 1533 for failing to provide the son he wanted.

Maria Sibylla Merian combined plants and the insects that fed upon them to great effect in her *Metamorphosis insectorum Surinamensium* (1705). Here the caterpillar of the iridescent blue *Morpho menelaus* butterfly strips the leaves of the pomegranate. Merian spent two years in Surinam, returning home with specimens and artwork to prepare her book.

TAB. V.

Enkelde Griet.
Octob. Nov.

Witte Platte Appel.
Sept. Octob.

Bloem - Suir.
Sept. Octob.

Brand-Appel.
Dec. Jan.

Heer - Appel.
Nov. Dec.

Eyer - Appel.
Oct. Novemb.

Rede Soete Jopen.
Octob. Novemb.

Pomme - Rose.
Oct. Nov.

Somer Striepeling.
Sept. Octob.

J. H. Knoop ad viv. del.

J. C. Philips sculpsit.

Apple

Malus domestica

Fruit of Temptation and Eternal Life

But of the tree of the knowledge of good and evil, thou shalt not eat of it: for in the day that thou eatest thereof thou shalt surely die.

Genesis 2:17

It seems fitting that if the apple's home cannot have been the Garden of Eden, it was at least in the 'celestial mountains'. Suggested locations for the biblical paradise don't easily provide the right conditions for sweet apples that can be plucked from the tree and eaten raw. They lack the essential period of sustained cold temperatures that stimulates the dormant buds to break and the blossom to appear when temperatures rise. By contrast, the slopes of the Tian Shan (which means 'celestial mountains'), in Central Asia, offered the necessary environment. The ancestors of today's table fruit most likely took root as part of a vast temperate forest that periodically stretched from the Atlantic to Beringia in the later Pliocene (up to 2.6 million years ago).

Today the slopes of the Tian Shan near Almaty (formerly Alma Ata) in Kazakhstan remain clothed in a veritable fruit forest (although it is under threat). Such abundance yields an array of tree forms, fruit sizes, colours, tastes, textures and ripening times in the native wild apple, *Malus sieversii*. This highly diverse species is the progenitor of our cultivated apple, *M. domestica*, which inherited the marvellous variation. Early cultivated apples spread from their Kazakh home along the Silk Roads, and the Greeks and Romans took them across Europe. It was in Europe that *M. domestica* met the local crab apple, *M. sylvestris*, and exchanged genes. Their hybrid progeny also exchanged genes with the parents, and over a long period of time these crosses (known as introgression) created the potential for our favourite apple varieties.

Once found, a variety can be maintained in only one way. Plant the pips from your favourite crunch and the next generation will be different. The five seeds in any one apple can give rise to five dissimilar plants. It is by grafting scion wood to a rootstock (first used with apples about 4,000 years ago) that the chance finds of Britain's perfect 'cooker' the 'Bramley's Seedling' (*c.* 1810s) or America's 'Newtown Pippin' (*c.* 1750s), 'Aport' or 'Alexander' from the Ukraine (*c.* 1700s), 'Gravenstein' (*c.* 1600s), which arose in Germany or Italy but became very popular around Hamburg in the 19th century, or Australia's world-beater, the 'Granny Smith' (*c.* 1860s) have been perpetuated.

Opposite A selection of apple varieties from Jean Herman Knoop, *Pomologie, ou Description des meilleures sortes de pommes et de poires* (1771). Note the natural imperfections on the fruit, often ignored in idealized portrayals.

Below Harvesting the ripe fruit, one of a series of woodcuts illustrating the annual cycle of maintaining a healthy orchard, in Marco Bussato's *Giardino di agricoltura* (1592). Bussato made his living from grafting before writing a successful series of books which attested to the growing interest in agronomy in the early modern period.

The doubt ecology has cast on the apple in Eden has been added to by philology. During the early years of the Christian Church in northern Europe an unwonted precision was attached to the more general Hebraic 'fruit' of Genesis 3:3. Similarly the Greek *melon* (or *malon*) could refer to any tree fruit: 'Armenian *melons*' were apricots, 'Persian *melons*', peaches, and 'Median *melons*', citrons. This offered multiple options for biblical scholars translating from the Greek and for those chasing the identity of the golden apples of the Garden of Hesperides, the securing of which formed the eleventh of Heracles' twelve labours. Hippomenes used three given him by Aphrodite to win a foot race against Atalanta, and thus secure her as his bride, when he dropped them and she stooped to pick them up. And if Paris had not been asked to judge which of three goddesses would receive the golden apple inscribed 'To the Fairest', the walls of Troy might have stood far longer.

Belief in the apple's sexual connotations, aphrodisiacal powers and place in courtship were widespread, and applied both to the larger, sweet ones and the small, sharp crab apples. Pre-Roman Celtic traditions venerated the latter. They made good cider and increased in palatability when dried or cooked. Devotees still make deliciously red crab apple jellies, the colour coming from tannins in the skin. As an ornamental tree, it is also increasingly appreciated.

Adam and Eve became mortal when they ate the apple, but Celtic and Norse myths imbued the fruit with powers of limitless longevity. The fabled King Arthur rests still in Avalon, where the apples of immortality grew. Scandinavia's magical elect lived by eating enchanted apples. When Idun, goddess of spring and keeper of the apples in the paradise of Asgard, is tricked into taking the fruits to their enemies the giants, the gods begin to shrivel and age. After various adventures she returns to her rightful place and restores youth and vigour to the gods with slices of apple.

The Romans were great planters of orchards. After the collapse of their empire, the custom was continued by the monastic orders in piecemeal fashion, although in Iberia it was the Muslims who restored pomological prosperity. Apples were grown for both the table and the cider press. Cider was an important drink throughout the apple-growing regions of Europe. And it was what many of Johnny Appleseed's trees were planted for in early 19th-century America. Part of a farm-labourer's wages were commonly paid in cider, a tradition that continued in Britain until 1878, when it was made illegal.

Dessert apples perhaps reached their apogee in Victorian Britain. Apples had had to compete with new delicacies after the voyages of exploration, but a combination of rising numbers of the newly moneyed building grand houses with walled kitchen gardens and the excellence of the British climate for raising apples fuelled a cult of the apple. The range of varieties and their extended ripening period reached new heights. The fruit was not just to be eaten, but also to be admired, the apple store duly opened for friends to exclaim over the contents.

Blossom and fruit of *Pyrus malus* (*Malus domestica*). Tian Shan, the home of the apple, was unscathed by the ice that scraped away much of the flora of northern Europe and North America, while considerable geological activity revitalized the soil and created new land. The region was also isolated by large desiccated areas surrounding its mountain ranges. All proved beneficial to the evolution of the apple.

PYRUS MALUS L.
Der Apfelbaum.

However, many were low yielding, biannual croppers, or liable to scabs, rots and insect pests. Elsewhere in the world, particularly in the USA, the tale of the 20th century was the rise of the large commercial orchard, industrialized picking, packing and sorting, and sale through the demanding supermarket chain. The number of varieties was radically pruned and new kinds were often bred for transportability over taste.

Today the value of forgotten varieties is being appreciated again, for what they offer to our palate and for the richness of their genetic makeup. Great wealth is to be found in the 'celestial mountains'. Our love affair with the apple is not over.

Chinese Plum or Japanese Apricot

Prunus mume

Herald of Spring

No. 179. *Toko-mume.*
Prunus mume var.
ROSACEÆ.

Flowers and fruit of 'Toko mume', a variety of *Prunus mume*, one of several that feature in *The Useful Plants of Japan* (1895). Japanese ume condiments include whole umeboshi pickles, a pureed ume paste or ume-su and the 'vinegar' left over from pickling.

The flowering plum is the earliest to blossom,
She alone has the gift of recognizing spring.

Xiao Gang, 6th century AD

China's mei trees have been cherished for thousands of years. Known as Chinese plum or Japanese apricot, this small tree is in fact closer to apricots than plums, their cousins among the *Prunus*. Native to the slopes of western Sichuan and western Yunnan in China, across the north of Vietnam and Laos, through Korea and Japan, its fragrant blossoms (*mei hua*), in whites, pinks, reds and pale greens, open on naked stems while snow still lies on the ground. Their appearance heralds the coming of spring. Together with the bamboo and pine, mei became known in China as the 'three friends in winter'.

Despite their beauty, mei were first cultivated for their yellow- or greenish-fleshed apricot-like fruits. Too sour to eat raw, they have been dried, salted, pickled, pureed and used to flavour wine. The contents of the Western Han dynasty tomb 1 at Mawangdui, Changsha (2nd century BC) included pots containing mei stones and dried fruit, as well as written evidence of processing techniques. Piquant pickles have been favourites in the East for millennia. Today's Chinese mei and Japanese ume condiments are widely eaten in East Asia and diasporas beyond. Belief in the fruit's medicinal qualities also has a long history, including as a wound salve and to give strength in battle. In combination with red perilla leaves (used as a colourant), pickled umeboshi's inherent chemicals – benzaldehyde and organic acids – have been shown to exhibit bactericidal action against *Escherichia coli*, particularly useful when consuming raw fish dishes.

The early orchards of the Han dynasty (206 BC–AD 200) were laid out for fruit production, but strikingly flowered cultivars attracted increasing attention. Poets of the 5th and 6th centuries began to extol their beauty and the five-petalled blossoms had great significance, being associated with the traditional five blessings of longevity, prosperity, health, virtue and a natural death. Writers celebrated the delicate yet robust nature of flowers that could survive the cold weather during Chinese New Year. Choice mei trees became highly desired in both imperial and private gardens. Their shapes were manipulated to produce preferred forms and vistas created with viewing pavilions. In time these would be replicated in miniature in the traditions of penjing in China and bonsai in Japan.

196

The Garden.

PRUNUS MUME
Pres. by W. Robinson, Esq. 7/97.

Branches in bloom were cut for indoor arrangements, although the flowers did not last long. Floral transience was juxtaposed with the tree's longevity. The blossoms, symbolizing rejuvenation and vigour, were borne on old wood, exemplifying steadfastness and durability. Artists depicted isolated, floating branches in exquisite detail. 'Ink plum blossom' (*mo-mei*) became a recognized genre.

During the Song (960–1279) and Yuan (1271–1368) dynasties planting of the trees, and artistic and poetic appreciation of the blossom, reached new heights and acquired new meanings. Song Po-jen's mid-13th century *Mei-hua hsi-shen p'u* (Register of Plum-blossom Portraits) contained 100 full-page woodcuts, each with an accompanying poem. It can be read as both a painting manual and a voice of protest against the Yuan Mongol rulers. The legacy of such artists – myriad decorative patterns and discontent expressed through art – continues to inspire, just as the blossoms themselves do each year, as winter hints of spring.

Watercolour of *Prunus mume*. Nanjing in China is home to the Purple Mountain Park where thousands of plum blossom trees scent the air. In Japan, festivals in February and March – *ume matsuri* – still celebrate these harbingers of spring in parks and shrine and temple grounds.

Rose

Rosa spp.

Flower of Love

white Rofe ; the efpeciall difference confi-
fleth in the colour and fmell of the floures;

Above 'Rosa Provincialis, sive
Damascena. The Province or
Damaske Rose' from Gerard's
Herball (1633). Gerard declared
the rose as deserving the 'chief
and prime place among all
floures [sic] whatsoever'.

Opposite 'Rosa bifera officinalis/
Rosier des Parfumeurs', from
P.-J. Redouté's Les Roses, vol.
1 (1817). Redouté's brilliant
execution and composition
captured almost everything
about the rose, except its
scent. The largest constituent
of rose oil is citronellol, but
it is other trace components
that contribute most to its
wonderful perfume.

*The red rose is part of the splendour of God; everyone who wants to look
into God's splendour should look at the rose.*

Ruzbihan Baqli of Shiraz, d. 1209

The rose is perhaps the most loved flower and one with many associations. One
hundred million roses changed hands in the week leading up to Valentine's Day
(14 February) 2013 at the world's largest flower auction, FloraHolland. The flower
has enjoyed periods of intense adoration, but never fallen from grace and *Rosa* continues to enthral in its many forms. From miniatures to sprawling groundcover,
through bush and shrub to climber and rambler; five-petalled, many-petalled;
whites, pinks, purples, reds, yellows, oranges, stripes, once and repeat flowering;
single stemmed and clustered blooms; evergreens and scented leaves; heavenly perfumes – the panorama of cultivated roses is a floral celebration like few others. It
attests to a passion to enjoy and manipulate this flower over many centuries, aided
by the almost uniquely complex way in which *Rosa*'s sets of seven chromosomes
separate and recombine.

The rose's symbolic meanings have moved with the times, mutating as fashions
and religions change, perhaps reflecting the fact that in many of the important
areas of ancient civilization there were local roses to cherish. *Rosa* is naturally confined to the northern hemisphere, but essentially encircled the globe between 20
and 70 degrees north. The genus appears to have thrived during the Oligocene
(33.9–23 million years ago), dispersing and diversifying into the approximately 100
to 150 species (rose taxonomy is complex) of wild roses gracing the planet today.

From these wild species the rose was probably domesticated many times.
The plant's innate ability to hybridize with close and far-flung relations when
brought into contact by human hands furnished the staggering number of garden
cultivars – around 20,000. Many older roses resulted from complex crosses, the
precise when and how remaining a mystery. The white and blush Albas are an old
European hybrid between *R. gallica* and *R. canina. R. damascena* has three parents,
each with a discrete natural range. At some point the species rose *R. gallica* and the
long-cultivated *R. moschata* produced a new hybrid. When this was crossed with
R. fedtschenkoana the glorious scented damascene or damask rose was born. There
is no ambiguity about its desirability: the damask is a perfume rose. Distillation
of its petals yields attar of roses and its by-product, rose water. Rose water is

41.

Rosa bifera officinalis. *Rosier des Parfumeurs.*

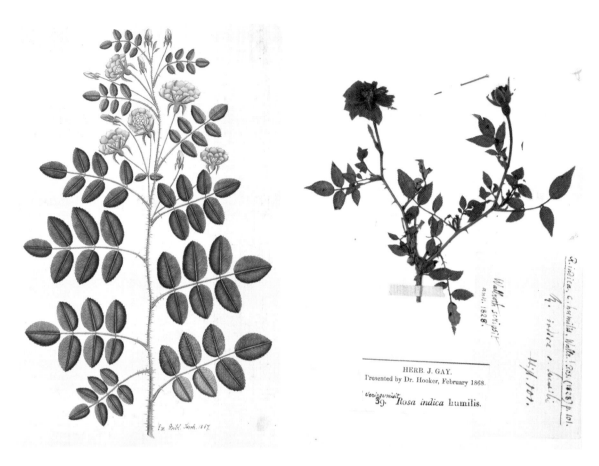

HERB. J. GAY.
Presented by Dr. Hooker, February 1868.
39. *Rosa indica* humilis.

Above left A watercolour probably painted in the early 19th century by one of the many Indian artists who produced botanical illustrations for the East India companies, combining traditional styles with the demands of European natural history illustration to produce images that were at once familiar and exotic.

Above right Herbarium specimen from the Royal Botanic Gardens at Kew of a variety of *Rosa indica* now known as *Rosa chinensis* Jacq. var. *minima*, presented by Joseph Dalton Hooker in 1868.

used routinely in sweet and savoury dishes of the Persian kitchen and aromatic confections throughout Southwest Asia. The Zoroastrians still greet guests with rose sweets.

The Greeks wrote sparingly of roses as emblems of death and love. The Romans continued these associations, but as with many things Roman, took them to new levels of excess. Roses featured in medicine and elite dining, and substantial cultivation such as in Roman Egypt and Campania produced the raw material for decorative garlands and chaplets. Demand was particularly high during the celebration of Rosalia, generally towards the end of May, when roses were placed on tombs in commemoration of the dead.

Initially shunned by the Christian Church, roses made a comeback in the Middle Ages, decorating churches and finding their way into spiritual writing. St Bernard of Clairvaux (1090–1153) in his exegesis of the Song of Songs lauded white roses as a Marian symbol, while the red rose's five petals frequently represented Christ's Five Wounds. The rose and cross were also combined in the Rosicrucians' secret society, which flourished in the swirling hermeneutics of 17th-century Protestantism.

Roses were a quintessential part of the Persian garden. In Islamic tradition it is said that as Muhammad ascended to receive his revelations, his sweat fell back to earth and a fragrant rose came forth. In the 13th century Sufi mystics such as Ruzbihan Baqli and Rumi celebrated the rose's transcendental qualities; not least the lingering ethereal perfume captured after the flower has faded.

By forcing or growing different kinds of roses it was possible to extend the rose season, but only the damask roses flowered twice in a year. In the late 18th century people in the West began to appreciate what the Chinese, with their vastly superior history of cultivation and different rose genes, had enjoyed for at least 1,000 years: remontant or repeat-flowering roses. The Chinese species bloomed all summer long in new shades of pink, crimson, blush and yellow, some with a delicate scent of fresh tea (from which came the hybrid teas). Four cultivars in particular survived the English climate (England having come to dominate the China trade) and served as 'studs', bringing their precious genes to fanciers in Europe and North America. Chinese roses started a breeding mania. Today's plants of enchanting beauty, scent and reliability are the results of the increasingly complex hybridizations (and some lucky sports). They delight enthusiasts in clubs and societies the world over.

Empress Josephine, wife of Napoleon, is often credited with playing a leading role in promoting the rose and establishing the concept of a dedicated rose garden at her chateau at Malmaison. Romantic yes, but not strictly accurate. While Josephine was certainly interested in roses she sprinkled them in glorious mixed borders; the 'rose garden' was invented when Malmaison opened to the public at the turn of the 20th century.

Roses can have a darker side too. Writing of the short reign (218–22) of the Roman emperor Elagabalus (Heliogabalus), Lampridius has him walking through rooms strewn with roses and suffocating his guests with masses of fragrant petals. The 'violets and other flowers' were turned into rose petals in Alma-Tadema's late 19th-century masterpiece *The Roses of Heliogabalus*. Today such implausible antics would probably require a shipment from Colombia, Ecuador or Kenya. They have captured the world market in these flowers, which have for so long captured our hearts.

Tulip

Tulipa spp.

A Mania for a Bulb

All these fools want is tulip bulbs/Heads and hearts have but one wish
Let's try and eat them; it will make us laugh/To Taste how bitter is that dish.

Petrus Hondius, 1621

Opposite A painting of *Tulipa greigii* by Mary Grierson (1912–2012), part of a series of tulips by one of the leading botanical artists of the later 20th century. Grierson worked as the herbarium artist at Kew, combining artistic skills with attention to detail and a fine understanding of botany.

Below A diminutive *Tulipa armena* (just over 20 cm/8 in. tall) from the Kew Herbarium. Small species tulips such as this were collected from the wild and manipulated (with the help of 'breaking viruses', which led to the stripes) into the fancy forms of the 17th-century tulip mania.

Tulip bulbs are humble looking things, easily held in the hand, a thin dry papery layer protecting the softer tissues within. But great potential lies deep inside. For here, in the centre of the scales that will provide its nourishment, is a new stem. At its tip is the embryonic flower bud. Once planted, and after the roots have formed and an essential period of coolness has passed, the stem will propel the growing flower bud upwards before its petals colour and open for a glorious two weeks or so of unbridled beauty.

In its quiescent state the tulip can easily be transported, ensuring instant gratification somewhere else the following spring. It was this that helped the Ottoman Turks transform various species of tulips from wild flowers to essential jewels of the garden. After his conquest of Constantinople (Istanbul) in 1453, Mehmed II built the Topkapı palace and laid out its Persian-inspired gardens, replete with carnations, roses, hyacinths, irises, jonquils and tulips. Under his successors, the tulip became intimately involved with Ottoman culture from the 16th century, evident in textiles, pottery and most famously its ornate glazed tiles.

Turkey may have been the place where the wild tulip entered the garden in grand style, but only four species of tulips are thought to be native here, although more now grow wild in mountainous areas of the country. The likely natural home of the tulip is further east, between the mountains of Tian Shan and Pamir-Alai in Central Asia. From here it spread where mountain and steppe provided the right combination of free drainage, winter chill, spring rain and a good baking in summer sun. The Ottoman sultans and their equally rapacious viziers were able to demand huge quantities of bulbs to be dug up from the wild in the empire's provinces and vassal states, while their florists began to manipulate nature. They selected and crossed the seeds of those tulips that grew ever closer to their ideals of perfection: long, slender flowers consisting of six same-length dagger-shaped petals, held upright on an elegant but strong stem.

For Europeans it wasn't quite love at first sight. The first tulip bulbs to arrive in Antwerp in 1562 suffered an inauspicious fate. Very much an adjunct to the bales of cloth he received, a merchant tried them roasted with oil and vinegar like

onions and tossed the rest into his garden. Most languished, but a few were treated with respect by George Rye, another merchant and keen gardener. Similar cargoes brought bulbs to Amsterdam. Some of Leiden's came when the new head of the botanic garden, Carolus Clusius, arrived in 1593. He hoarded his blooms, refusing to share or sell, and fell victim to theft. Dutch tulipomania is the most famous, but France and England experienced periods of intense frenzy, and countries such as Germany were influential markets.

Tulip-fancying was part of the wider interest in exotic plants brought back through journeys of trade and exploration, itself part of the passion for owning and displaying natural curiosities. Plants, and especially tulips as tulipomania took hold, were to be grown in precisely arranged beds of exact dimensions, which visitors were invited to witness and admire as the flowers reached their peak.

And what a peak it was. European taste in tulips favoured a plumper look to the flower; the flower heads were often large, on long stems. But what drove tulip madness was the colour. Not the pure, solid colours familiar in today's mass plantings and flower-shop blooms, but the delicate feathering and flaming of tulips that by some occult process had become multicoloured. This was termed 'breaking' – one colour apparently breaking into another on the petals as if painted with the finest brush. There were three desired types. Bizarres were red streaked on yellow, Roses red on white and Bybloemens purple on white.

Such astonishing beauty is now known to be caused by viruses, which interfere with the genes that code for proteins producing particular pigments in the petals. The virus also weakens the plant and slows the formation of bulblets. These baby bulbs were grown on, bulking up the stock (seeds do not come true). Over time what is essentially a diseased individual plant dies out, but its progeny continue. There is also the potential for new favourites as parasitic virus and tulip host co-existed in the florists' beds. So, ironically, it was damaged goods that were traded with such passion initially by aficionados and then in the 1620s by the emergent professional nurserymen. Such businesses profited as the price of bulbs (sold by weight after lifting) rose and they continued to prosper as entrepreneurial burgomasters entered the market as middlemen trading in tulip futures. This was the bubble that burst in late 1636 and early 1637.

The Dutch organized successful packaging for long-distance transport and used travelling salesman to create markets for their wares. Such enterprise would see them become the world's leading growers of bulbs and flowers, moving into the massive American market with new breeds of pure coloured flowers in the 20th century. Today the Dutch grow 4.32 billion tulips, 2.3 billion of which fill flower stalls and bouquets as cut flowers. They are now as virus-free as possible and those flamboyant blooms of the past live on only among a few cognoscenti growers and in the still-coveted still life paintings of the old masters.

Opposite A plate from Robert Thornton's *Temple of Flora, or Garden of Nature* (1799–1807). Thornton employed leading artists and engravers, and oversaw the production of the illustrations. He had the plants set against naturalistic backgrounds and the effect was stunning, but the project bankrupted him and he died in poverty.

Orchids
Orchidaceae
Strange and Beautiful Blooms

An orchid in a deep forest sends out its fragrance even if no one is around to appreciate it. Likewise, men of noble character hold firm to their high principles, undeterred by poverty.

Confucius, 551–479 BC

Above The 'Angurek katong'ging,' first described by Engelbert Kaempfer in *Amoenitatum Exoticarum* (1712), combining the Malay words for epiphytic orchid and scorpion; it is now known as *Arachnis flosaeris*. *Arachnis* species were important in commercial breeding of hybrid orchids for the houseplant market.

Opposite The *Cymbidium hookerianum* orchid, found in Bhutan in 1848 and named by Heinrich Gustave Reichenbach for Joseph Dalton Hooker. Reichenbach was one of the leading orchidists of the 19th century. The artist and engraver Walter Hood Fitch was a master at fitting large, complex plants into paper sizes needed for publication.

Orchids have it all: they are exquisite, fragrant, alluring, and come in a seemingly unending diversity of extraordinary floral forms that enchant and engage the mind. If orchids are eternally beautiful, today they are also big business. Women may not wear orchid corsages as they once did and orchids must vie with roses and lilies in bridal bouquets, but step into a supermarket and there are potted orchids by the shelf-full. Many are *Phalaenopsis* or 'moth orchids' – named for their resemblance to those denizens of the insect world. Choice abounds among these unnamed mass-market hybrids bred from species from South Asia and the Indonesian archipelago and raised using micro-propagation techniques. They may offend the orchid purists, but the whites, pinks, purples and yellows, and spots and stripes of the often scented flowers have a grace and elegance that bring an aura of the exotic, a sense of the humid tropics to the windowsill. Today's multibillion dollar business is but the latest growth of an ancient fascination. Wherever they occur – over huge swathes of the planet with the exception of the frozen Antarctic – orchids have been held in high esteem.

Confucius associated orchids with purity and morality and they later became a favourite theme for the literati who retired from public life rather than serve the conquering Mongols of the Yuan dynasty. In poetry and especially in ink drawings, orchids symbolized high-mindedness in adversity. Orchid painting would develop into a leading art form in China, Japan and Korea. The plants were also employed in the Chinese pharmacopeia. Various species of *Dendrobium* were renowned for their strengthening powers, properties that reflected this orchid's challenging habitat. Named Shih-hu or 'rock-living', it was thought that plants that could thrive when clinging to exposed rock must be resilient and robust and would therefore impart these qualities. Together with other Chinese orchids they do contain active substances which are currently being investigated.

A similar association between plant forms and medicinal use was known in ancient Greece and Rome as the 'doctrine of signatures'. Europe is home to only about 1 per cent of the world's orchid species, including members of the *Orchis* (Greek for testicles) and *Ophrys* genera, with paired, subterranean, swollen *rchid-* X574

Cfr. C. giganteum ... *Cymbidium Hookerianum. C grandiflora* ...

The hybrid orchid *Cypripedium morganiae* (*Paphiopedilum morganiae*) from the Scrapbooks of John Day (1824–88), who painted his own orchids, those in London nurseries and the Royal Botanic Gardens, Kew, and some in situ in the tropics.

rhizomes, thought to resemble male generative organs. In Greek mythology Orchis enjoyed various sexual adventures but these ended badly, and transformed after death he bequeathed the world these sensuous plants. Theophrastus (*c.* 372–288 BC), Dioscorides (*c.* AD 40–80) and Galen (AD 129–*c.* 210) linked assorted orchids with fecundity and suggested their use in fertility disorders. Various species have recently been identified on Roman monuments, appearing as part of the frieze on the Ara Pacis or peace altar built by Augustus and on the Temple to Venus Genetrix dedicated by Julius Caesar. Folk traditions continue to recommend these plants as aphrodisiacs. The Turks have long enjoyed an ice cream made of salep – the dried powdered root of *Orchis mascula* – and make claims for its positive effects.

The Orchidaceae family is one of the largest among the flowering plants, with in excess of 26,000 species, most found in the tropics and subtropics. Many are epiphytes – air plants that grow high up in the canopy. In the 17th century, voyages of discovery in search of East Indies spices brought botanizing merchants, missionaries, officers, diplomats and doctors into new territories, with their exotic floras. Georg Eberhard Rumphius, 'first merchant' of Batavia (Jakarta), produced a twelve-volume *Herbarium Amboinense* (published posthumously in the mid-18th century). He included important new species of terrestrial and epiphytic orchids, the latter described as 'aristocrats of wild plants, who convey their nobility by wanting to live only high up in trees'.

Orchis mascula, the early purple orchid, which, like *O. morio* was used by Darwin in his research. The twinned swollen rhizomes inspiring the genus name are displayed, along with the seed capsules, each of which can contain 4,000 seeds. Destruction of their habitat (old meadows and pastures) means that even this prodigality cannot contend with 21st-century life.

Such books and the arrival of the plants themselves fuelled the new passion for exotics. Sir Joseph Banks brought dendrobium orchids back from Australia. Humbler servants of the expanding British Empire also returned home with specimens although many perished on the long sea voyages. The survivors could be kept alive in hothouses, but the mistaken view that they were parasitic plants led to fatal mistakes in their culture and much disappointment as they faded or failed to flower. William Cattley, an exotic-plant enthusiast, succeeded. The 'most splendid' *Cattleya labiata* from the Organ Mountains of Brazil displayed its true beauty in his hothouse in 1818. Cattleyas would become extremely popular.

Nurseries had initially relied upon informal networks of botanical enthusiasts as suppliers of parent stock, but they began to use professional plant hunters. In the 19th century, nurserymen such as James Veitch gambled on the heavy costs involved and sent out men to brave the rigours of the tropics. As orchid mania intensified in Britain and northern Europe in the second half of the century the effects on the natural habitats could be devastating. Competing collectors stripped orchids by the thousands, taking all they could find in a locality, occasionally by felling the host trees to pick off the plants at ground level. Once they had taken all they could carry they destroyed the rest to prevent others gaining access to them.

To bulk out the collected stocks – many of which were auctioned on arrival – nurseries also experimented with propagation and hybridization. Orchids often failed to set seeds for several years, and when they did, germination was no easy matter. Orchids produce copious dust-like seeds. More air than embryo, they float in the air or perhaps on water. Miniaturization means they lack the food stores that feed larger seeds as they germinate; to compensate, orchids evolved species-specific relationships with mycorrhizal fungi to aid germination and subsequently support the plant in continued symbiosis. Still unaware of this, the persistent and talented plantsman John Dominy, working for Veitch, raised the first flowering

TAB.11.   GENERA.

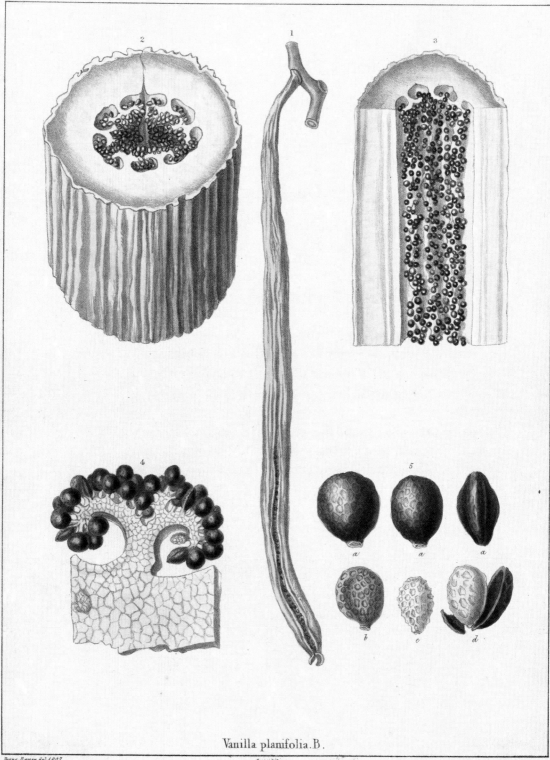

Vanilla planifolia. B.

Franz Bauer del. 1807.   Gauci lith.g   Printed by C. Hullmandel.

hybrid orchid *Calanthe × dominyi* in 1856. It was named in his honour by the leading orchidist John Lindley, who remarked to its creator: 'You will drive the botanists mad'.

Charles Darwin retained his sanity but was fascinated with orchids and their means of reproduction. He wanted to understand the amazing array of adaptations that have co-evolved between different species of orchids and their individual insect pollinators. Working first with native British orchids, many then growing around his home at Down House in Kent, he described orchids that imitated female insects, inducing the males to try to mate but instead picking up pollen from one plant and fertilizing the next. He was also well supplied with tropical orchids from his friends and correspondents. The stunning white orchid from Madagascar, *Angraecum sesquipedale*, with its 25-cm (10-in.) long nectary amazed him. His suggestion of a long-tongued moth as pollinator would prove correct. Identified in 1903, *Xanthopan morganii praedicta* has a tongue that can indeed reach the nectar and pick up the pollen along the way, but it wasn't actually caught in the act until 1997.

The special relationship between plants and pollinators played a part in the one orchid that has perhaps universal acclaim. Very few people do not like vanilla, with its complex, aromatic, floral flavour. But what they may not know is that the real thing, so much better than the synthetic substitute usually made from lignin, comes from orchids. The main vanilla orchid is *Vanilla planifolia* (or *fragrans*). Cherished by the Totonacs of eastern Mexico, its fermented beans were used to pay tribute to the Aztecs and flavoured the cold chocolate drink reserved for the elite or taken by soldiers before engagements. The Totonacs revered the orchid, believing that the first plant grew from the blood of an ancient princess whose punishment for eloping with her lover was to be beheaded.

Vanilla is still expensive today because outside its homeland (with its natural pollinator) it has to be hand-pollinated, and then handpicked and hand-cured before being hand-sorted and packed. The discovery in 1841 of a simple method of hand-pollination by a young, recently freed slave, Edmond Albius, on Réunion in the Indian Ocean, allowed this island to overtake Mexico as the world's largest producer in the 1860s. Word of Albius's method and the vanilla orchid spread, so that today Madagascar and Indonesia – both home to many beautiful indigenous orchids – are the leading producers.

Orchids still fascinate and surprise. In 2011 *Bulbophyllum nocturnum*, newly discovered on New Britain (Papua New Guinea), was seen opening its flowers at 10 p.m. – the first night-flowering orchid ever observed. Botanist André Schuiteman at Kew predicts that the pollinator will be a tiny fungus gnat since the orchid's flowers resemble slime moulds. Such is the wonder of the orchid world.

Opposite 'The parts of the fruit of *Vanilla planifolia*', the vanilla orchid, with the seeds magnified x 200. The botanist and orchid specialist John Lindley and renowned artist Francis Bauer teamed up to produce *Illustrations of Orchidaceous Plants* (1830–38), which has been described as a foundation work on orchids.

Peony

Paeonia spp.

Flower of Riches and Honour

In front of the Emperor's audience hall many thousand-petalled tree peonies were planted ... he would sigh and say 'Surely there has never before been such a flower among mortals'.

9th-century Chinese writer

'Pivoine odorante', from P.-J. Redouté's *Choix des plus belles fleurs* (1827–33). Moutan or tree peonies, *P. suffruticosa*, originated in China, where they had long been avidly cultivated. Held in high esteem by the Tang dynasty emperors, their massive blooms were used to ornament the imperial palaces. They were still relatively new and very sought after when Redouté painted them early in the 19th century.

In some Greek legends, Paeon (or Paean) served as physician to the gods. He cured Hades and Ares of their wounds and as a reward after his death was transformed into a peony. The story could relate to one or more of five species of herbaceous peonies that graced the hillsides of Greece and the Greek islands. Paiawon is mentioned in a list of deities on a Linear B tablet found at Knossos, Crete, so perhaps it was the pure white single-petalled Cretan *P. clusii* that is referred to. This plant's roots have high levels of the volatile antimicrobial paeonol. Deep maroon *P. parnassica* found on Mount Parnassus contains methyl salicylate (similar to aspirin), so external use could have helped aches and sprains.

But neither of these species came to dominate the learned herbals. In the 1st century AD Dioscorides gave various medical applications for peonies and referred to two kinds – male and female. This related not to the plant's reproductive strategy, which was not then appreciated (and in any case peonies are hermaphrodite), but to their relative robustness. *P. mascula* (the Balkan peony) has been identified as male and *P. officinalis*, female. 'She' came to dominate the medical market, and found favour as a hardy garden plant. In medieval Europe peonies were part of the floral complement of a 'noble garden'. Crushed seeds were used as spice and baked roots eaten alongside roast pork, their astringency a perfect foil for fatty meat. Double whites and reds, with their showy, petal-rich flower bowls were bred: the peony aesthetic was well underway.

Herbaceous peonies reach from Spain to Japan, from the Kola Peninsula on the Arctic Circle to Morocco (there are also two further herbaceous species in northwestern America and Mexico). China is fortunate to be home to the eight species of tree peonies too. Mudan or Moutan (often *P. suffruticosa*) peonies are taller, woody plants that have gained their own mystique and glamour. Their name is said to come from Mudang – the mythical Emperor of Flowers. During the Tang dynasty (618–907) tree peonies became a veritable mania at court and beyond, and inspired poets and artists. Some exchanged hands for a hundred ounces of gold and gardeners achieved blooms of 30 cm (12 in.) in diameter. Also native to China, the herbaceous *P. lactiflora* was grown en masse as a medicinal.

Puaonia daurica.

Buddhist monks took the tree peony to Japan in the 8th century. Here gardeners developed the imperial peonies and later the anemone-flowered forms, manipulating the stamens (male flower parts) until they resembled narrow petals and appeared to fill the petal bowl. They gave the resulting plants elaborate names.

After their introduction by Sir Joseph Banks at the end of the 18th century, tree peonies became a sensation in Europe's elite gardens – a scented, thorn-free 'rose'. And when *P. officinalis* met *P. lactiflora* their offspring were considered great beauties. Plant hunters sought out new species in the East for introduction and breeding; this was not straightforward since China's borders were difficult to penetrate, but persistence paid off. So too did that of Japan's Toichi Itoh, who finally achieved his determined desire for a double yellow by overcoming the previously unfruitful crossing of tree and herbaceous parents. Sadly, he died before his plants came into their floral glory in 1963 and showed the way to yet more new forms and colours in this glorious flower.

Watercolour of '*Paeonia daurica*' – this herbaceous peony ranges from Croatia to Iran, across Turkey and the Caucasus mountains. In the past there was some uncertainty about its relationship to *P. mascula*, the 'male' peony of Dioscorides' great herbal from the 1st century AD, but recent work has confirmed it as a separate species.

Wonders of Nature
The Extraordinary Plant World

Above Detail of a watercolour of a red waterlily, painted in India by Mrs Fanny C. Russell between 1860 and 1870.

Is it invidious to privilege some plants more than others, because they fascinate, or astound, perhaps even repel? Yet there are plants that have inspired great wonderment, and even the hardened (and hardy) souls familiar with the world's most biodiverse regions have been brought to a standstill by certain plants. Nature's plant wonders are not merely sensory splendours. Scratch the surface, and their adaptations help reveal the history of life on Earth, in all its abundance and tenacity.

The peculiar upside-down or baobab trees bestride the thirsty regions of Africa, Madagascar and Australia. They serve as giant water towers and indigenous people literally tap the swollen trunk for its stored water. Welwitschias cope with their limited water supply in a range of ways, from their ever-growing leaves to a special kind of metabolism. Squat and low, welwitschias sit out the seasonal drought and sandstorms of their home in the Namib Desert. So far, they have managed to survive as a species from the age of the dinosaurs, when life was warmer and wetter.

If the welwitschia is renowned for it ugliness, few would dispute the loveliness of the giant waterlily native to the waterways of South America. While the pineapple-scented flowers attract the attention of pollinating beetles, the structure of the leaves – a masterclass in vegetable engineering – inspired the ironwork for London's Crystal Palace, home to the Great Exhibition of 1851. Evolution has led

to some near optimum configurations in the natural world. The special arrangement of the seeds in a sunflower head and the plant's initial ability to track the sun both inspire solar-engineering projects.

Pitcher plants and the giant rafflesia find the nutrients they need in singular ways. Pitchers grow in nitrogen-poor areas and compensate for this deficiency by catching insects (and sometimes larger prey) in their highly modified leaves. The pitcher holds a solution of acidic digestive enzymes. Must-have plants of the 19th-century hothouse, the exploitation in the wild reached dangerous levels. Rafflesias are the world's largest single flowers. They are also parasites, living on the roots of tropical vines and attracting their pollinators, carrion flies, by emitting the smell of rotting meat. Loss of the host vine through habitat destruction in the forests will also mean that these exceptional plants will vanish.

Man-made habitat destruction doesn't come much harsher than the atomic bomb dropped on Hiroshima, Japan, at the close of the Second World War. Less than a mile from the epicentre of the bomb a ginkgo tree rose like a phoenix from its burnt remains. Our relationship with the plant world is ancient, and plants do seemingly limitless things for us, but it is still a work in progress. Utility aside, a plant simply being there, part of the web of life, warrants a profound celebration.

Above left The baobab tree's curious pendant flower, in a watercolour perhaps drawn from the specimens growing in the Botanic Garden, Calcutta, around the turn of the 19th century.

Above right Details of a *Welwitschia mirabilis*, including (bottom centre) the insertion of the leaves, as well as the cones, scales, flower parts and seeds that make up the reproductive structures of this amazing plant.

Baobab

Adansonia spp.

The Upside-Down Tree

They carried me to a particular spot where I saw a herd of antelopes; but I laid aside all thoughts of sport, as soon as I perceived a tree of prodigious thickness, which drew my whole attention.

Michel Adanson, 1759

The flower and leaf of the baobab, with the fruit behind, from *Curtis's Botanical Magazine* (1828). The fruits were said to be traded by the Mandingo people of West Africa. Damaged fruit were burnt and the ash boiled with rancid palm oil to make soap. Dried leaves were mixed with food and used medicinally.

According to one African creation myth, God gave each animal its own tree. The hyena was the last to receive his and got the baobab. He was so disgusted that he threw the tree away and it landed upside down; the baobab's distinctive shape does make its branches look like a root system. It has always been a botanical curiosity to outside observers, but it was the complete tree for those who lived in the areas of its unusual geographical distribution.

The genus *Adansonia* is named after the French naturalist Michel Adanson, who first encountered the baobab around 1750. But the tree had been known in Egypt and the Middle East for centuries, prized especially for its fruit – mixed with water it produces a refreshing drink. This fruit, sold in Cairo markets, was understood by a Venetian naturalist in the late 16th century to be called *bu hobab*, although this may have been *bu hibab*, 'the fruit with many seeds'. Locally, the whole tree could be turned to use: the leaves and flowers were eaten as salad, the seeds were roasted like coffee beans, the pliable bark was pounded into rope or its fibres were woven into cloth, and the hard outer shell of the fruit made excellent dishes or containers. The enormous trunk of older trees could store great quantities of water, which was then available during the dry season. The trees were occasionally fitted with a tap for ease of drawing. The huge size and hollow centres mean that trunks have been turned to a variety of uses, including a pub, a prison and a restaurant.

The trees' vast size – as much as 30 m (100 ft) in circumference – gives the genus greater girth (if not total volume) than the giant redwoods of California, and early naturalists assumed that baobabs were amazingly long-lived. Adanson reasoned from the fact that two trees inscribed by 15th- and 16th-century travellers had grown so little between then and his visit that the trees must be upwards of 5,000 years old. The explorer-missionary David Livingstone admired the trees, but didn't think they had survived Noah's Flood. Modern estimates give the tree a maximum lifespan of about 2,000 years, although precise determination is difficult.

Baobab's natural distribution is also unusual. The genus consists of eight species: the African one (*Adansonia digitata*) is the most common. It prefers relatively dry

savannah conditions, about 450–600 m (1,475–1,970 ft) above sea level. There is also a single Australian species (*A. gibbosa*), indigenous to the Kimberley region of Western Australia. The other six species are all found in Madagascar, where baobabs almost certainly originated. It is thought that perhaps a million years ago hard-coated fruit floated the relatively short distance from Madagascar to the African coast, and then made the much longer journey to Australia. Time, and different environmental conditions, led to the two new species.

Its unusual size and shape, combined with the many uses to which the tree could be put, endowed the baobab with special meanings to native peoples. Trees could have a spiritual significance, and doubled as places of worship. Many have individual names and when they die have been accorded full funeral rites. An ability to grow new trunks from fallen branches means that a single tree can extend over a large area. Although specimen trees have been grown in many tropical countries, at least two of Madagascar's six species are now threatened because of land clearing and neglect.

Baobab near the Bank of the Lue (1858), an oil painting by Thomas Baines which captures the tremendous girth of this group of trees on a tributary of the Zambezi River. The baobabs fascinated European travellers not least for the use of their hollow trunks as burial chambers for African bards or griots. These people were refused burial for fear that their involvement with the occult during life might contaminate the soil or water, and in the dry heat they became mummified.

Welwitschia

Welwitschia mirabilis

Strange Desert Phenomenon

Out of all question the most wonderful plant ever brought to this country – and the very ugliest.

Joseph Dalton Hooker, 1862

The Namib, one of the world's oldest deserts, runs down the Atlantic coastal plain of southwest Africa. Rainfall is meagre and erratic. Coastal fogs bring vital water to the plant and animal life, drip by condensed drip. On a small strip in the north of this desert region lives the 'ugliest' plant in the world. It may not be beautiful, but the *Welwitschia mirabilis* is certainly extraordinary.

Sunk into the earth is a deep, water-seeking taproot. Near its top a network of spongy roots reaches out to extract any hint of groundwater from the infrequently wetted watercourses the plants inhabit. Above this stands a short, partly submerged woody stem, with a dished top. Each side of the stem gives rise to a single leaf, though occasional plants have more than one set of leaves. The leaves continue to grow indefinitely from the base, increasing in length 10 to 15 cm (4 to 6 in.) each year for at least 600 years. Lose a leaf and the plant dies. Over such long periods sandstorms rag the leaves into ribbons, giving the illusion of more than the actual two. They provide occasional food for grazing animals, both large – antelopes, rhino – and small – tiny insects known as microarthropods.

Deserts make enormous demands on their flora, and it is odd to see such large leaves in this environment. Welwitschias are better than most plants at reflecting solar radiation, but they lack some of the usual leaf adaptations of desert dwellers – small size, specialized water storage organs and waxy surfaces. What they have evolved to do is practice flexibility in the way their stomata – the tiny pores that allow water and gases in and out – open and close, and in the storage of carbon (from carbon dioxide) as organic acids ready for photosynthesis.

Welwitschia is certainly an ancient plant, tracing its direct ancestors back beyond 200 million years ago when seed-bearing plants had first gained dominance. Fossils indicate that the Welwitschia family were once much more widespread than their current circumscribed homeland, and lived in much moister conditions than today. While they may have thrived as the continents of Africa and South America separated and outlived the extinction of the dinosaurs, welwitschias have only been part of modern botanical science since the 1860s when reports

and material from two explorers reached the centre of the botanical world: the Royal Botanic Gardens, Kew, in London.

The plant is named for Friedrich Welwitsch, an Austrian-born doctor, botanist and explorer who went collecting for the Portuguese government and found his strange plant in the southernmost part of their colony of Angola. Although the plant immediately attracted the attention of the botanical community, his bosses were disappointed. Welwitschia might be a unique plant, but it had no economic importance.

The English artist and explorer Thomas Baines trekked into the Namib and sent material to Kew, where Joseph Dalton Hooker spent many arduous hours at the microscope, producing an outstanding book on this peculiar plant. Hard work this may have been, but it didn't compare with Welwitsch's suffering in his pursuit of plants. As was all too common among explorers, he endured malaria, dysentery, scurvy and badly ulcerated legs. All this for a plant 'that he could do nothing but kneel down in the burning soil and gaze at … half in fear lest a touch should prove it a figment of the imagination'.

A younger (top) and more mature specimen (bottom) of *Welwitschia mirabilis* (not to the same scale). The leaves of the older plant have the characteristic shredded appearance of a long life in the desert's winds.

Giant Waterlily

Victoria amazonica

'A Vegetable Wonder'

No words can describe the grandeur and beauty of the plant.

Joseph Paxton to the Duke of Devonshire, 2 November 1849

For Robert Schomburgk finding the 'vegetable wonder' was particularly special. His exploration of Guyana's rivers on behalf of Britain's Royal Geographical Society had been fraught, but on 1 January 1837, in a shallow basin of the River Berbice, he came upon the massive leaves (up to 2.5 m or 8 ft across) of the giant waterlily floating on the still water. Immense white and pink flowers (30 cm/12 in. across) were supported on thick, prickly stalks, and as night fell Schomburgk became aware of their delicious pineapple-tinged scent. He wasn't the first European to see this plant (that happened in Bolivia in 1801), but its rediscovery was celebrated. Seeds sent by Schomburgk were successfully germinated by William Hooker, director at Kew Gardens, and the plants would be the first to flower outside South America.

The giant waterlily, eventually named *Victoria amazonica* in honour of the British queen, caused a sensation when Joseph Paxton, head gardener to the Duke of Devonshire at Chatsworth, succeeded in bringing the plant he'd received from Kew into bloom in November 1849. Paxton had to design a dedicated glasshouse for the waterlily's heated tanks, such was the plant's typical rate of growth (15 cm or 6 in. a day) and need for warmth. This multi-talented man would go on to produce the blueprint for the Great Exhibition's Crystal Palace in 1851. For both the waterlily house and the Crystal Palace, Paxton drew inspiration from nature – in fact from the underside of the waterlily's leaf, a 'natural engineering feat'.

Readers of the *Illustrated London News* were already familiar with the whimsical image of Paxton's daughter Annie, her weight distributed on a tin tray, standing on a waterlily's leaf. As he would explain at a meeting of the Society of Arts – with a piece of leaf as a prop – he had exploited the plant's innate load-bearing ability, with cantilevers stretching from the leaf's middle to the edge and the connecting system of stout but flexible ribs and cross braces, for his innovative 'ridge and furrow' roof design.

The flower is a 48-hour marvel. Opening white, scented and heated by a thermochemical reaction on its first evening, the female flower attracts its pollinating beetle, already laden with pollen. The flower subsequently closes, trapping the

beetle, which eats, stays warm and fertilizes the flower. Then, changing from white to pink, female to male, and lacking warmth and scent, the flower opens again the next evening, releasing the beetle. Now covered in fresh pollen, the insect moves on in search of more fragrant white warmth. The fertilized flower closes for the final time and disappears under the water, job done.

This beetle–waterlily synergy may have been happening for a very long time. Fossils of *Microvictoria svitkoana* dating from some 90 million years ago have been placed with *Victoria* in the family Nymphaeaceae. There is such correspondence between the structure of the living flowers and those of the fossils (found in an old clay pit at Sayreville, New Jersey, USA) that these ancient blooms must also have been insect pollinated. What is different is the size. The *Microvictoria* are minute, just 1.6 mm (¹⁄₁₆ in.) in diameter. This ancient lineage can tell us much about the emergence of the angiosperms, or flowering plants, in the history of life. Schomburgk might well have been pleased with this new, fossilized 'vegetable wonder'.

The stunning flower and a section of the equally impressive leaf of the giant waterlily. Many botanists have subsequently shared Joseph Paxton's careful study of the leaf's architecture to understand what prevents it from distorting in agitated water. The immense surface also needs to be able to drain water. The slightly depressed ribs channel surface water to two drainage sinuses and water is also shed through special pores.

Pitcher Plant

Nepenthes spp.

Caught in a Trap

I may here mention a few of the more striking vegetable productions of Borneo. The wonderful Pitcher-plants ... here reach their greatest development.

Alfred Russel Wallace, 1869

As a general rule, insects (and other animals) eat plants, but some plants strike back. Around 650 species of plants (and a few fungi) attract and digest animals, mostly insects but occasionally larger prey. This seems so contrary to the symmetry of the natural world that the great naturalist Linnaeus refused to believe it, even though many examples had been seen and illustrated. Charles Darwin, whose classic book of 1875 on the subject made them better known to the world, called such plants 'insectivorous', but their even greater exploitation of animal products means that 'carnivorous' is now the preferred term.

Carnivorous plants, which are widely distributed and belong to several botanical groups, use a variety of active and passive mechanisms to trap their suppers. They often inhabit wet areas, although a few are adapted to drier climates. Some, including the Venus Flytrap (*Dionaea muscipula* – Darwin called it 'the most wonderful plant in the world'), produce sticky excretions that attract their prey, after which the trap snaps shut on them. Others have attractive scents. Pitcher plants, named from their remarkable leaf modifications shaped like jugs or pitchers, are passive and simply rely on insects or other animals, attracted by colour, scent or nectar, falling into the liquid contained within. The pitchers vary widely in colour, shape and size; some can hold as much as 2 litres (4 pints), and rats have been seen in them.

The liquid inside the pitcher, although often diluted with rainwater, is naturally slightly acidic. Many species have a flap over the top to limit the amount of rainfall that collects. On the inside of the pitcher is an area called the 'pitfall zone', which has a waxy secretion. When touched by the intruder, surface crystals drop off, making the inner coat extremely slippery. When something is trapped, the acidity of the liquid increases dramatically, and this aids digestive enzymes in the breakdown of the tissues of the victim, from which the plant absorbs nitrogen and other nutrients. The pitcher's walls are thin but extremely tough.

Asian pitcher plants belong to the genus *Nepenthes*. It contains about 110 species that are distributed throughout tropical regions from India to northern Australia, with the highest concentration in the Malay Archipelago. Although it

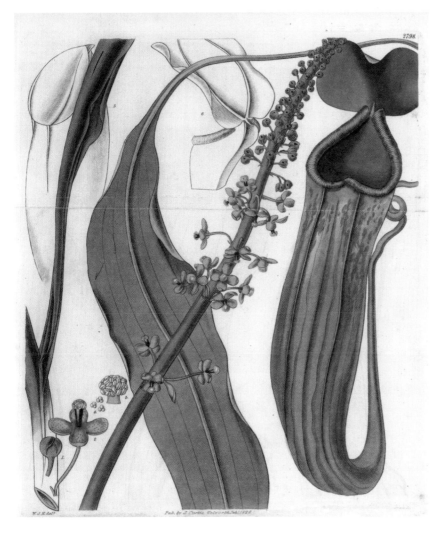

2798.

is a large genus, individual species generally have quite small ranges, and a few have been found just once, so are known only from their type specimen. However, many species are very vigorous and some grow well on wasteland. Most are vines, and can be as long as 20 m (66 ft); a few are shrubs. The range of adaptations is astonishing, with some growing out of tree trunks and others in very dry, stony ground.

Serious European fascination began in the late 18th century, when specimens were shipped back, and during the Victorian period many were grown in hothouses. The challenge to gardeners was to reproduce the appropriate growing conditions for a particular species, depending on its natural habitat. Trade in all *Nepenthes* species is now controlled.

The development of a 'pitcher' to trap animals also exists in a family of New World plants, which include the Cobra Lily and several species of the genus *Sarracenia*. In a case of tit-for-tat, mosquitoes that are resistant to the digestive enzymes of one species use the fluid in the pitcher as a breeding pool.

Nepenthes distillatoria, the native pitcher plant of Sri Lanka, was one of the earliest recorded in the botanical literature, in the later 17th century. Its stems were traditionally twisted and used as cattle rope in Sri Lanka.

Rafflesia
Rafflesia arnoldii
The Biggest Bloom

Buds of *Rafflesia patma*
increase in size after emerging
from their host's roots over a
protracted period. They can be
vulnerable in their forest habitat
– mammals such as porcupines
and tree shrews eat them,
crows damage them looking for
insects and a range of animals
trample them.

Come with me, Sir come! a flower, very large, beautiful, wonderful!

Joseph Arnold, letter to Dawson Turner, Pulau Lebar, on the Manna River,
Sumatra, May 1818

On 19 May 1818, Dr Joseph Arnold botanizing in the rainforests of southern Sumatra met with what he judged must be the world's largest flower. At up to 1 m (3 ft) in diameter and 7 kg (15 lb) in weight, *Rafflesia arnoldii* has maintained its place at the top of the big flower league. But size is just one of its many fascinating attributes.

Arnold, a naval surgeon, had been befriended by Sir Stamford Raffles, himself a keen amateur naturalist, who in 1817 had been appointed lieutenant-governor at Benkulen (Bengkulu) on the west coast of Sumatra. Raffles was delighted when Arnold accepted the post of naturalist and it was during a field trip they made in the rainforest, two days upriver from the coast, that a servant ran to bring news of this 'greatest prodigy of the vegetable world'. In just over five weeks Arnold would be dead of fever, probably malaria. The plant was later named *Rafflesia arnoldii* in honour of both men.

Arnold was not the first European to see a *Rafflesia*. That honour belongs to Louis August Deschamps, who discovered a smaller species in 1797, subsequently called *Rafflesia patma*. Some 19 species have now been identified, dotted across the islands of Sumatra, Java, Borneo, the Philippines and on the Malay/Thai peninsula. They are rare plants, threatened by habitat destruction as the rainforests are cleared for timber or land.

All *Rafflesias* are parasites. They have no leaves, roots or stem; and with no chlorophyll they do not photosynthesize. Most of their life is in fact spent as thread-like filaments hidden within the tissues of their host plant, a few species of the vine *Tetrastigma* (which is in the grape family). The strange flowers emerge through the bark of the vine's roots as tightly closed buds that grow to resemble cabbage heads, before five, thick, warty-looking petals unfold. *Rafflesia* have separate male and female flowers and the pollen is carried from one to the other by carrion flies, for these short-lived giants (in bloom for five days) resemble a rotting carcass and have a scent to match, hence their popular name, the corpse flower. In return for their aid in pollination the flies receive no reward: *Rafflesia* are doubly parasitic.

Carrion mimicry is frequently associated with floral gigantism and may be connected in part with why these super blooms evolved. During a rapid expansion in the size of *Rafflesia* flowers there was also a switch to pollination by carrion flies or beetles (sapromyophily). In a kind of arms race, the bigger the flower the more attractive it is and so the more often it is visited, particularly if large carrion are preferred. It may also have been a way for different species of *Rafflesia* to co-exist in the same locale without risk of hybridization. Different sizes make the mechanics of pollination very difficult and *Rafflesia* species range from 10 cm (4 in.) to the giant *arnoldii*.

There are no fossils of these plants. They emerged some 46 million years ago and in that time have also experienced extensive gene transfers from their host plants. If their habitat is destroyed and they become extinct, the chance to understand more of what has driven their extraordinary evolution will be lost forever.

The corpse flower, *Rafflesia arnoldii*, has the world's largest individual flower but neither roots nor leaves. A parasite, it relies upon a host, a type of vine found in Sumatra and Borneo. The flower bud bursts through the bark, opens, grows and emits the stench of putrefying flesh to attract carrion flies for pollination.

RAFFLESIA ARNOLDI R.BROWN

Sunflower

Helianthus annuus

Nature's Inspiration

… in one summer, beeing sowne of a seed in Aprill, it hath risen up to the height of fourteene foot … one floure was in weight three pound and two ounces … and sixteen inches broad.

John Gerard, 1636

The sunflower's prodigious ability to capture the sun's energy for rapid growth could be said to have brought out a competitive streak. One of the features herbalists in 16th-century Europe reported about this new plant from the Americas was the considerable height gained in its single growing season. The current record stands at 8.23 m (27 ft), for a sunflower grown in Kaarst, Germany.

Whether or not the people of eastern North America vied to grow the tallest, they had been cultivating sunflowers for a long time before the plant reached the Old World. Evidence points to domestication between 5,000 and 4,500 years ago. Large, single-headed sunflowers were part of the early crop complex, serving as an important source of edible oil, as the plant still does. Wild sunflowers also continued to be collected, eaten and used in medical and ceremonial contexts in the western states.

Early trade routes probably introduced domesticated sunflowers to Mexico. Archaeological evidence certainly hints at a broad and ancient use beyond the table. Sunflowers were part of Aztec ceremonies connected with sun worship. The Mexican flowers were yellow-centred – it's not hard to see the association between this and the life-giving sun. After the Spanish conquest the Catholic clergy sought to stamp out such adulation.

Sunflower heads are made up of numerous small, tubular disc florets packed together in the centre of the flower and surrounded by ray florets, which resemble a ring of petals. The disc florets form intricate, intersecting spirals, an effect easier to see after each floret has produced its 'seed' or achene (in fact a fruit containing a seed). These patterns are known as Fibonacci spirals: the number of achenes in each spiral follows the sequence 1, 1, 2, 3, 5, 8 (each number is the sum of the previous two), and each seed is turned through the golden angle of around 137 degrees. This is the most economical way to pack in the possible 1,000-plus achenes in a flower head, so that all of them can be the same size and with no crowding at the centre or wasted space at the edges. The pattern may help to improve the spacing of mirrors in concentrated solar power plants to reflect the sun's rays towards a central electricity-generating tower, while reducing wasteful shadows.

'Portrait de l'herbe du soleil de Monardes', in Claude Duret's *Histoire admirable des plantes et herbes …* (1605). Nicolas Monardes never travelled to the New World, but his botanical garden in Seville was stocked with South American plants and his descriptions of the sunflower were among the earliest in Europe.

Helianthus gigantius Lin: sp. pl.

Opened sunflower heads don't track the sun (heliotropism); they mostly face eastwards as if to catch the sunrise. But the actively growing parts of the flower – unopened buds, with their green bracts and young leaves – do. Special cells near the base of the bud and leaf stalk regulate water pressure, facilitating directional growth. The green parts are the plant's powerhouses. If they can maximize the uptake of solar energy, and water, carbon dioxide and essential nutrients are also abundant, growth will be optimized. The sunflower can be used directly for biomass energy production, but perhaps its greatest contribution will lie in the inspiration it provides to solar engineers. Photovoltaic cells mounted so they can move and follow the sun are potentially more efficient than fixed arrays at generating electricity. But movement costs energy. Engineers have developed a flexible support system that utilizes the sun's heat to tilt the cells, overcoming this problem. Patents are pending: sunflower sun worship may be back in fashion.

The sunflower in this watercolour by an unknown Indian artist in the Company style is the epitome of a solar disc. There is evidence that sunflowers can play a role in phytoremediation – using plants to decontaminate soil and, if grown hydroponically, water.

Ginkgo
Ginkgo biloba
The Great Survivor

This Tree's leaf, which, from the Orient, Is entrusted to my garden,
Lets us savour a secret meaning, As to how it edifies the learned man.

J. W. von Goethe, 1815

The unmistakable leaves of the ginkgo in Engelbert Kaempfer's *Amoenitatum Exoticarum* (1712). Kaempfer was the first outsider to describe the tree and he probably drew this illustration himself. There are still trees alive in the region of Nagasaki that would have been growing when he visited.

The ginkgo is ancient: its ancestors could have been grazed by dinosaurs, so old is the tree. Its distinctive foliage, with its bilobed leaves (hence its species name), has been preserved in the fossil record dating back about 200 million years. Several aspects of the ginkgo tree's mode of reproduction and internal structure also highlight its antiquity, making it a living laboratory of evolution.

Individual ginkgo trees can live several hundred years: the oldest in England, in Kew Gardens, was planted by 1762, when George III was on the throne. Many specimens in China, Korea and Japan, where the tree is revered, are even older – a few are said to have lived for several thousand years, though these ages are probably legendary. Some are possibly more than 1,000 years old, however, and the Grand Ginkgo King, in Li Jiawan, southern China, is about 30 m (100 ft) tall and sports a trunk more than 5 m (16 ft) across. Many of the old giants are found at shrines and sites of pilgrimage. In oriental cultures the tree is associated with longevity, and its products – leaves as well as seeds – are used medicinally for a variety of complaints, including memory loss and urinary problems. The nuts are also consumed at the table and ginkgo tea is sometimes drunk.

Millions of years ago, the ginkgo was widely spread across the world, including much of North America. But it failed to adapt to colder climates, and perhaps it lost whatever animals aided in the fertilization of the separate male and female trees. There are two remote forests in China where the stands may be original, but ginkgos have been venerated there for so long that humans may have planted even these ancient forests. More certain is that the tree was transported to Japan and Korea centuries ago and planted in gardens, temples and shrines. Legend says that Confucius sat under ginkgo trees, reading, meditating and teaching.

It was in Japan that the German naturalist and doctor Engelbert Kaempfer described and named the tree in 1691, from his transliteration of its Japanese name. He was intrigued by the biological peculiarities of the tree's reproduction and the putrid smell of the fruit produced by females. This smell (likened to fresh vomit) is so powerful that male specimens, which produce no fruit, were favoured for planting in the West. Seeds and young trees proved adaptable to a wide variety of climates and situations, even polluted cities. The tree is a familiar sight in New York, Beijing and many other modern urban landscapes, in streets and parks. From the brink of total extinction a thousand or so years ago, the ginkgo is now safe, and in a reversal of the usual pattern this is through human agency.

The ginkgo's autumnal leaves are stunning in the weeks before they rapidly fall. It has been doing this for millions of years, which makes it a great survivor. Nowhere is this more clearly demonstrated than by six ginkgos of Hiroshima: two years after the atomic bomb dropped on 6 August 1945, they sent up shoots from their charred remains. One was less than a mile from the bomb's epicentre.

Robert Fortune commissioned a series of drawings of trees during his last visit to China in the 1850s, including *Salisburia adiantifolia* as the ginkgo was then known. Painted by an unknown watercolourist, each tree was accompanied by a human figure – the artist only agreed to undertake the commission if he could add his people.

Further Reading

General

Arber, Agnes, *Herbals: Their Origin and Evolution* (Cambridge University Press, Cambridge, 1912; repr. The Lost Library, n.d.)

Beerling, David, *The Emerald Planet: How Plants have Changed Earth's History* (Oxford University Press, Oxford & New York, 2007)

Blunt, Wilfrid, *The Art of Botanical Illustration* (3rd ed., Collins, London, 1955)

Blunt, Wilfrid and S. Raphael, *The Illustrated Herbal* (Frances Lincoln, London, 1994)

Campbell-Culver, Maggie, *The Origin of Plants* (Headline, London, 2001)

Davidson, Alan, *The Oxford Companion to Food* (Oxford University Press, Oxford & New York, 1999)

Fry, Carolyn, *The Plant Hunters* (Andre Deutsch, London, 2012)

Goody, Jack, *The Culture of Flowers* (Cambridge University Press, Cambridge & New York, 1993)

Grove, A. T. and Oliver Rackham, *The Nature of Mediterranean Europe* (Yale University Press, New Haven, 2001)

Hora, Bayard (ed.), *The Oxford Encyclopedia of Trees of the World* (Oxford University Press, Oxford & New York, 1981)

Kingsbury, Noël, *Hybrid: The History and Science of Plant Breeding* (University of Chicago Press, Chicago, 2009)

Kiple, Kenneth F., *A Movable Feast: Two Millennia of Food Globalization* (Cambridge University Press, Cambridge, 2007)

Kiple, Kenneth F. and Kriemhild Conceé Ornelas (eds), *The Cambridge World History of Food*, 2 vols (Cambridge University Press, Cambridge & New York, 2000)

Knapp, Sandra, *Potted Histories: An Artistic Voyage through Plant Exploration* (Scriptum, London and Firefly Books, Buffalo, NY, 2003)

Murphy, Denis J., *People, Genes and Plants: The Story of Crops and Humanity* (Oxford University Press, Oxford & New York, 2007)

Musgrave, Toby, Chris Gardner and Will Musgrave, *The Plant Hunters* (Cassell, London, 1999)

Musgrave, Will and Toby Musgrave, *An Empire of Plants: People and Plants that Changed the World* (Cassell, London, 2000)

Newton, John, *The Roots of Civilization: Plants that Changed the World* (Pier 9, Millers Point, NSW, 2009)

North, Marianne, *A Vision of Eden: The Life and Work of Marianne North* (Holt, Rinehart and Winston, New York, 1980)

Prance, Ghillean and Mark Nesbitt (eds), *The Cultural History of Plants* (Routledge, New York & London, 2005)

Radkau, Joachim, *Wood: A History*, trans. Patrick Camiller (Polity, Cambridge, 2012)

Sherwood, Shirley and Martyn Rix, *Treasures of Botanical Art* (Kew Publishing, London, 2008)

Silvertown, Jonathan, *Demons in Eden: The Paradox of Plant Diversity* (University of Chicago Press, Chicago, 2005)

Toussaint-Samat, Maguelonne, *History of Food*, trans. Anthea Bell (2nd ed., Wiley-Blackwell, Chichester and Malden, MA, 2009)

Walker, Timothy, *Plants: A Very Short Introduction* (Oxford University Press, Oxford & New York, 2012)

Transformers

Albala, Ken, *Beans: A History* (Berg, New York, 2007)

Alexander, Caroline, *The Bounty: The True Story of the Mutiny on the Bounty* (Viking, New York and HarperCollins, London, 2003)

Besnard, G. et al., 'The complex history of the olive tree: from Late Quaternary diversification of Mediterranean lineages to primary domestication in the northern Levant', *Proceedings of the Royal Society B* (2013), 280: 2012–33

Coe, Sophie D., *America's First Cuisines* (University of Texas Press, Austin, 1994)

Cohen Suarez, Amanda and Jeremy James George, *Handbook to Life in the Inca World* (Facts on File, New York, 2011)

Coursey, David, *Yams: an account of the nature, origins, cultivation and utilisation of the useful members of the Dioscoreaceae* (Longmans, London, 1967)

DuBois, C. M., Chee-Beng Tan and S. Mintz (eds), *The World of Soy* (University of Illinois Press, Champaign, 2008)

Fuller, D. Q. and E. L. Harvey, 'The archaeobotany of Indian pulses: identification, processing and evidence for cultivation', *Environmental Archaeology* 11 (2006), 219–46

Fuller, D. Q., 'Pathways to Asian civilisation: tracing the origins and spread of rice and rice cultures', *Rice* 4(3) (2011), 78–92

Kaniewski, David et al., 'Primary domestication and early uses of the emblematic olive tree: palaeobotanical, historical and molecular evidence from the Middle East', *Biological Reviews* 87(4) (2012), 885–99

Kessler, D. and P. Temin, 'The organization of the grain trade in the early Roman Empire', *Economic History Review* 60(2) (2007), 313–32

Lee, Gyoung-Ah et al., 'Archaeological Soybean (*Glycine max*) in East Asia: Does Size Matter?', *PLoS ONE* 6(11) (2011), e26720

Lotito, Silvina B. and Balz Frei, 'Consumption of flavonoid-rich foods and increased plasma antioxidant capacity in humans: cause, consequence, or epiphenomenon?', *Free Radical Biology & Medicine* 41(12) (2006), 1727–46

Loumou, Angeliki and Christina Giourga, 'Olive groves: "The life and identity of the Mediterranean"', *Agriculture and Human Values* 20(1) (2003), 87–95

McGovern, Patrick E., *Ancient Wine: The Search for the Origins of Viniculture* (Princeton University Press, Princeton, 2003)

McGovern, Patrick E., *Uncorking the Past: the Quest for Wine, Beer, and other Alcoholic Beverages* (University of California Press, Berkeley, 2009)

Mann, Charles C., *1491: New Revelations of the Americas Before Columbus* (Vintage, New York, 2006)

Meyer, Rachel S., Ashley E. DuVal and Helen R. Jensen, 'Patterns and processes in crop domestication: an historical review and quantitative analysis of 203 global food crops', *New Phytologist* 196(1) (2012), 29–48

Molina, Jeanmaire et al., 'Molecular evidence for a single evolutionary origin of domesticated rice', *Proceedings of the National Academy of Sciences* 108(20) (2011), 8351–56

Mueller, Tom, *Extra Virginity: The Sublime and Scandalous World of Olive Oil* (W. W. Norton, New York, 2011)

Muller, M. H. et al., 'Inferences from mitochondrial DNA patterns on the domestication history of alfalfa (*Medicago sativa*)', *Molecular Ecology* 12(8) (2003), 2187–89

Myles, Sean et al., 'Genetic structure and domestication history of the grape', *Proceedings of the National Academy of Sciences* 108(9) (2011), 3530–35

Phillips, Rod, *A Short History of Wine* (Allen Lane, London, and Ecco, New York, 2000)

Reader, John, *Propitious Esculent: The Potato in World History* (William Heinemann, London, and Yale University Press, New Haven, 2009)

Salavert, Aurélie, 'Olive cultivation and oil production in Palestine during the early Bronze Age (3500–2000 BC): the case of Tel Yarmouth, Israel', *Vegetation History and Archaeobotany* (2008) 17/Supplement 1, 53–61

Staller, John E., *Maize Cobs and Cultures: History of Zea mays L.* (Springer, New York, 2009)

Various, 'The Origins of Agriculture: New Data, New Ideas', *Current Anthropology* (Wenner-Gren Symposium Supplement 4) Vol. 52 (October) 2011

Various, 'From collecting to cultivation: transitions to a production economy in the Near East', *Vegetation History and Archaeobotany* (special issue), Vol. 21 (2), 2012

Zeven, A. C. and W. A. Brandenburg, 'Use of paintings from the 16th to 19th centuries to study the history of domesticated plants', *Economic Botany* 40(4) (1986), 397–408

Zhang, Gengyun et al., 'Genome sequence of foxtail millet (*Setaria italica*) provides insights into grass evolution and biofuel potential', *Nature Biotechnology* 30 (2012), 549–54

Zohary, Daniel, Maria Hopf and Ehud Weiss, *Domestication of Plants in the Old World* (4th ed., Oxford University Press, Oxford & New York, 2012)

Taste

Albala, Ken, *Eating Right in the Renaissance* (University of California Press, Berkeley, 2002)

Block, Eric, *Garlic and Other Alliums* (Royal Society of Chemistry, Cambridge, 2010)

Brown, Pete, *Hops and Glory: One Man's Search for the Beer that Built the British Empire* (Pan, London, 2010)

Cornell, Martyn, *Beer: The Story of the Pint: The History of Britain's Most Popular Drink* (Headline, London, 2003)

Cox, D. N. et al., 'Acceptance of health-promoting Brassica vegetables: the influence of taste perception, information and attitudes', *Public Health Nutrition* 15(8) (2012), 1474–82

Dalby, A., *Dangerous Tastes: The Story of Spices* (British Museum Press, London, and University of California Press, Berkeley, 2000)

Fritsch R. and N. Friesen, 'Evolution, domestication, and taxonomy', in H. D. Rabinowitch and L. Currah (eds), *Allium Crop Science – Recent Advances* (CAB International Publishing, Wallingford, 2012), 5–30

Hornsey, Ian S., *A History of Beer and Brewing* (Royal Society of Chemistry, Cambridge, 2003)

Keay, John, *The Spice Route* (John Murray, London, 2005, and University of California Press, Berkeley, 2006)

Li, Hui-Lin, 'The vegetables of ancient China', *Economic Botany*, 23(3) (1969), 253–60

Livarda, Alexandra, 'Spicing up life in north-western Europe: exotic food plant imports in the Roman and medieval world', *Vegetation History and Archaeobotany*, 20 (2011), 143–64

Milton, Giles, *Nathaniel's Nutmeg: How One Man's Courage Changed the Course of History* (Farrar, Strauss and Giroux, New York, 1999)

Mitchell, S. C., 'Food idiosyncrasies: beetroot and asparagus', *Drug Metabolism and Disposition* 29(4) (2001), 539–43

National Onion Association, *Onions – Phytochemical and Health Properties*, http:// onions-usa.org/img/site_specific/uploads/ phytochemical_brochure.pdf

Ninomiya, Kumiko, 'Umami: a universal taste', *Food Reviews International* 18(1) (2002), 23–38

Pelchat, M. L. et al., 'Excretion and perception of a characteristic odor in urine after asparagus ingestion: a psychophysical and genetic study', *Chemical Senses* 36(1) (2011), 9–17

Schier, Volker, 'Probing the mystery of the use of saffron in medieval nunneries', *The Senses & Society* 5(1) (2010), 57–72

Yilmaz, Emin, 'The chemistry of fresh tomato flavor', *Turkish Journal of Agriculture and Forestry* 25 (2001), 149–55

Zanoli, Paola and Manuela Zavatti, 'Pharmacognostic and pharmacological profile of *Humulus lupulus* L.', *Journal of Ethnopharmacology* 116(3) (2008), 383–96

Heal and Harm

Baumeister, A. A., M. F. Hawkins and S. M. Uzelac, 'The myth of reserpine-induced depression: role in the historical development of the monoamine hypothesis', *Journal of the Neurosciences* 12(2) (2003), 207–20

Buckingham, John, *Bitter Nemesis: The Intimate History of Strychnine* (CRC Press, Boca Raton, 2007)

Che-Chia, C., 'Origins of a misunderstanding: the Qianlong Emperor's embargo on rhubarb exports to Russia, the scenario and its consequences', *Asian Medicine: Tradition and Modernity* 1(2) (2005), 335–54

Cousins, S. R. and E. T. F. Witkowski, 'African aloe ecology: A review', *Journal of Arid Environments* 85 (2012), 1–17

Davenport-Hines, Richard, *The Pursuit of Oblivion. A Global History of Narcotics, 1500–2000* (Weidenfeld & Nicolson, London, 2001)

Desai, P. N., 'Traditional knowledge and intellectual property protection: past and future', *Science and Public Policy* 34(3) (2007), 185–97

Duffin, J., 'Poisoning the spindle: serendipity and discovery of the anti-tumor properties of the vinca alkaloids', *Canadian Bulletin of Medical History*, 17(1) (2000), 155–92

Foust, C. M., *Rhubarb: The Wondrous Drug* (Princeton University Press, Princeton, 1992)

Hefferon, Kathleen, *Let Thy Food Be Thy Medicine: Plants and Modern Medicine* (Oxford University Press, Oxford & New York, 2012)

Hodge, W. H., 'The drug aloes of commerce, with special reference to the Cape species', *Economic Botany* 7(2) (1953), 99–129

Honigsbaum, Mark, *The Fever Trail: The Hunt for the Cure for Malaria* (Macmillan, London, 2001)

Hsu, Elisabeth, 'The history of *qing hao* in the

Chinese *materia medica*', *Transactions of the Royal Society of Tropical Medicine and Hygiene* 100(6) (2006), 505–08

Jay, Mike, *High Society: Mind Altering Drugs in History and Culture* (Thames & Hudson, London, and Park Street Press, Rochester, VT, 2010)

Jeffreys, Diarmuid, *Aspirin: The Remarkable Story of a Wonder Drug* (Bloomsbury, London & New York, 2004)

Laszlo, Pierre, *Citrus: A History* (University of Chicago Press, Chicago, 2007)

Laveaga, Gabriela Soto, 'Uncommon trajectories: steroid hormones, Mexican peasants, and the search for a wild yam', *Studies in History and Philosophy of Biological and Biomedical Sciences* 36(4) (2005), 743–60

Marks, Lara, *Sexual Chemistry: A History of the Contraceptive Pill* (Yale University Press, New Haven & London, 2010)

Maude, Richard J. et al., 'Artemisinin antimalarials: preserving the "Magic Bullet"', *Drug Development Research* 71(1) (2010), 12–19

Miller, Louis H. and Xinzhuan Su, 'Artemisinin: discovery from the Chinese herbal garden', *Cell* 146(6) (2011), 855–58

Neimark, Benjamin, 'Green grabbing at the 'pharm' gate: rosy periwinkle production in southern Madagascar', *The Journal of Peasant Studies* 39(2) (2012), 423–45

Newsholme, Christopher, *Willows: The Genus 'Salix'* (Batsford, London, 2002)

O'Brien, C., et al., 'Physical and chemical characteristics of *Aloe ferox* leaf gel', *South African Journal of Botany* 77 (2011), 988–95

Rocco, Fiammetta, *The Miraculous Fever-Tree: Malaria, Medicine and the Cure that Changed the World* (HarperCollins, London & New York, 2003)

Smith, Gideon F. and Estrela Figueiredo, 'Did the Romans grow succulents in Iberia?', *Cactus and Succulent Journal* 84(1) (2012), 33–40

Sun, Yongshuai et al., 'Rapid radiation of *Rheum* (Polygonaceae) and parallel evolution of morphological traits', *Molecular Phylogenetics and Evolution* 63(1) (2012), 150–58

Wright, C. W. et al., 'Ancient Chinese methods are remarkably effective for the preparation of artemisinin-rich extracts of qing hao with potent antimalarial activity', *Molecules* 15(2) (2010), 804–12

Woodson, Robert E. et al., *Rauwolfia: Botany, Pharmacognosy, Chemistry & Pharmacology* (Little, Brown, Boston, 1957)

Technology and Power

Barber, Elizabeth Wayland, *Women's Work: The First 20,000 Years; Women, Cloth, and Society in Early Times* (W.W. Norton, New York, 1995)

Bass, George F., *Beneath the Seven Seas* (Thames & Hudson, London & New York, 2005)

Borougerdi, Bradley J., 'Crossing conventional borders: introducing the legacy of hemp into the Atlantic world', *Traversea* 1 (2011), 5–12

Curry, Anne, *Agincourt: A New History* (Tempus, Stroud, 2010)

Farjon, Aljos, *A Natural History of Conifers* (Timber Press, Portland, 2008)

Friedel, Robert, *A Culture of Improvement: Technology and the Western Millennium* (MIT Press, Cambridge, MA, 2007)

Fu, Yong-Bi et al., 'Locus-specific view of flax domestication history', *Ecology and Evolution* 2(1) (2011), 139–52

Goodman, Jordan and Vivien Walsh, *The Story of Taxol: Nature and Politics in the Pursuit of an Anti-cancer Drug* (Cambridge University Press, Cambridge & New York, 2001)

Green, Harvey, *Wood: Craft, Culture, History* (Viking, New York, 2007)

Hageneder, Fred, *Yew: A History* (Sutton, Stroud, 2011)

Hardy, Robert, *The Longbow: A Social and Military History*, (5th ed., J. H. Haynes and Co., Yeovil, 2012)

Isagi, Y. et al., 'Clonal structure and flowering traits of a bamboo [*Phyllostachys pubescens* (Mazel) Ohwi] stand grown from a simultaneous flowering as revealed by AFLP analysis', *Molecular Ecology* 13(7) (2004), 2017–21

Kelchner, Scot A., 'Higher level phylogenetic relationships within bamboos (Poaceae: Bambusoideae) based on five plastid markers', *Molecular Phylogenetics and Evolution* 67(2) (2013), 404–13

Meiggs, Russell, *Trees and Timber of the Ancient World* (Clarendon Press, Oxford, 1982)

Renvoize, Stephen, 'From fishing poles and ski sticks to vegetables and paper: the bamboo genus *Phyllostachys*', *Curtis's Botanical Magazine* 12(1) (1995), 8–15

Tudge, Colin, *The Secret Lives of Trees* (Allen Lane, London & New York, 2006)

Woolmer, M., *Ancient Phoenicia: An Introduction* (Bristol Classical Press, London, 2011)

Yafa, Stephen, *Cotton: The Biography of a Revolutionary Fiber* (Penguin, London, 2005)

Young, Peter, *Oak* (Reaktion Books, London, 2013)

Cash Crops

Abbott, Elizabeth, *Sugar: A Bittersweet History* (Duckworth Overlook, London, 2009)

Balfour-Paul, Jenny, *Indigo: Egyptian Mummies to Blue Jeans* (British Museum Press, London, 2011)

Coe, Sophie D. and Michael D. Coe, *The True History of Chocolate* (3rd ed., Thames & Hudson, London & New York, 2013)

Corley, R. H. V and P. B. H. Tinker, *The Oil Palm* (4th ed., Blackwell Science, Oxford and Malden, MA, 2003)

Davies, Peter, *Fyffes and the Banana* (Athlone Press, London, 1990)

Ellis, M., *The Coffee House: A Cultural History* (Weidenfeld & Nicolson, London, 2004)

Goodman, Jordan, *Tobacco in History: The Cultures of Dependence* (Routledge, London & New York, 1994)

Grandin, Greg, *Fordlandia: The Rise and Fall of Henry Ford's Forgotten Jungle City* (Metropolitan Books, New York, 2009, and Icon Books, London, 2010)

Hobhouse, H., *Seeds of Change: Five Plants that Transformed Mankind* (Sidgwick & Jackson, 1985, and Harper & Row, New York, 1986)

Legrand, Catherine, *Indigo: The Colour that Changed the World* (Thames & Hudson, London & New York, 2013)

Loadman, John, *Tears of the Tree: The Story of Rubber – A Modern Marvel* (Oxford University Press, Oxford & New York, 2014)

Mair, Victor H. and Erling Hoh, *The True History of Tea* (Thames & Hudson, London & New York, 2009)

Mann, Charles C., *1493: How Europe's Discovery of the Americas Revolutionized Trade, Ecology and Life on Earth / Uncovering the New World Columbus Created* (Granta, London, and Knopf, New York, 2011)

Moxham, Roy, *Tea: Addiction, Exploitation and Empire* (Constable, London, and Carroll & Graf, New York, 2003)

Pendergrast, Mark, *Uncommon Grounds: The History of Coffee and how it Transformed the World* (Basic Books, New York, 2010)

Tan, K. T. et al., 'Palm oil: Addressing issues and towards sustainable development', *Renewable & Sustainable Energy Reviews* 13(2) (2009), 420–27

Landscape

Arakaki, Mónica et al., 'Contemporaneous and recent radiations of the world's major succulent plant lineages', *Proceedings of the National Academy of Sciences* 108(20) (2011), 8379–84

Brownsey, Patrick, *New Zealand Ferns and Allied Plants* (2nd ed., David Bateman, Auckland, 2000)

Campbell-Culver, Maggie, *A Passion for Trees: The Legacy of John Evelyn* (Eden Project Books, London, 2006)

Hora, Bayard (ed.), *The Oxford Encyclopedia of Trees of the World* (Oxford University Press, Oxford, 1981)

Johnstone, James A. and Todd E. Dawson, 'Climatic context and ecological implications of summer fog decline in the coast redwood region', *Proceedings of the National Academy of Sciences* 107(10) (2010), 4533–38

Large, M. F. and J. F. Braggins, *Tree Ferns* (Timber Press, Portland, 2004)

McAuliffe, J. R. and T. R. Van Devender, 'A 22,000-year record of vegetation change in the north-central Sonoran Desert', *Palaeogeography, Palaeoclimatology, Palaeoecology* 141(3) (1998), 253–75

Preston, Richard, *The Wild Trees: What if the Last Wilderness is Above our Heads?* (Allen Lane, London, 2007)

Prytherch, David L., 'Selling the eco-entrepreneurial city: natural wonders and urban stratagems in Tucson, Arizona', *Urban Geography* 23(8) (2002), 771–93

Rackham, Oliver, *The Last Forest: The Fascinating Account of Britain's Most Ancient Forest* (Dent, London, 1993)

Thomas, Graham Stuart, *Trees in the Landscape* (Frances Lincoln, London, 2004)

Tuck, Chan Hung et al., 'Mapping mangroves', *Tropical Forest Update* 21(2) (2012), 1–26

Williams, Cameron B. and Stephen C. Sillett, 'Epiphyte communities on redwood (*Sequoia sempervirens*) in northwestern California', *The Bryologist* 110(3) (2007), 420–52

Wolf, B. O. and C. Martinez del Rio, 'How important are columnar cacti as sources of water and nutrients for desert consumers? A review', *Isotopes in Environmental and Health Studies* 39(1) (2003), 53–67

Revered and Adored

Allan, Mea, *Darwin and his Flowers: The Key to Natural Selection* (Faber, London, 1977)

Arditti, Joseph et al., '"Good Heavens what insect can suck it" – Charles Darwin, *Angraecum sesquipedale* and *Xanthopan morganii praedicta*', *Botanical Journal of the Linnean Society* 169(3) (2012), 403–32

Avanzini, Alessandra (ed.), *Profumi d'Arabia* (L'Erma di Bretschneider, Rome, 1997)

Berliocchi, Luigi, *The Orchid in Lore and Legend*, trans. Lenore Rosenberg and Anita Weston (Timber Press, Portland, 2000)

Bennett, Matt, 'The pomegranate: marker of cyclical time, seeds of eternity', *International Journal of Humanities and Social Science* 1(19) (2011), 52–59

Bickford, Maggie, *Bones of Jade, Soul of Ice* (Yale University Art Gallery, New Haven, 1985)

Bickford, Maggie, 'Stirring the pot of state: the Southern Song picture book *Mei-Hua Hsi-Shen P'u* and its implications for Yuan scholar-painting', *Asia Major* 3rd series, 6(2) (1993), 169–236

Bickford, M., *Ink Plum: The Making of a Chinese Scholar-Painting Genre* (Cambridge University Press, Cambridge & New York, 1996)

Browning, Frank, *Apples. The Story of the Fruit of Temptation* (Penguin, London, 1998)

Chadwick, A. A. and Arthur E. Chadwick, *The Classic Cattleyas* (Timber Press, Portland, 2006)

Cobb, Matthew Adam, 'The reception and consumption of eastern goods in Roman society', *Greece and Rome* (Second Series) 60(1) (2013), 136–52

Cribb, Phillip and Michael Tibbs, *A Very Victorian Passion: The Orchid Paintings of John Day* (Thames & Hudson, London, 2004)

Dash, Mike, *Tulipomania* (Gollancz, London, and Crown Publishers, New York, 1999)

Ecott, Tim, *Vanilla: Travels in Search of the Ice Cream Orchid* (Grove Press, New York, 2004)

Fay, Michael F. and Mark W. Chase, 'Orchid biology: from Linnaeus via Darwin to the 21st century', *Annals of Botany* 104(3) (2009), 359–64

Fearnley-Whittingstall, Jane, *Peonies: The Imperial Flower* (Weidenfeld & Nicolson, London, 1999)

Fisher, John, *The Origins of Garden Plants* (Constable, London, 1989)

Garber, Peter M., 'Tulipmania', *Journal of Political Economy* 97(3) (1989), 535–60

Griffiths, Mark, *The Lotus Quest: In Search of the Sacred Flower* (Chatto & Windus, London, 2009, and St. Martin's Press, New York, 2010)

Hansen, Eric, *Orchid Fever: A Horticultural Tale of Love, Lust and Lunacy* (Methuen, London, and Pantheon Books, New York, 2000)

Hsü, Ginger Cheng-Chi, 'Incarnations of the Blossoming Plum', *Ars Orientalis* 26 (1996), 23–45

Ji, LiJing et al., 'The genetic diversity of *Paeonia* L.', *Scientia Horticulturae* 43 (2012), 62–74

Johnston, Hope, 'Catherine of Aragon's Pomegranate, revisited', *Transactions of the Cambridge Bibliographical Society* 13(2) (2005), 153–73

Juniper, Barrie E., 'The mysterious origin of the sweet apple: on its way to a grocery counter near you, this delicious fruit traversed continents and mastered coevolution', *American Scientist* 95(1) (2007), 44–51

Juniper, B. E. and D. J. Mabberley, *The Story of the Apple* (Timber Press, Portland, 2006)

Nikiforova, Svetlana V. et al., 'Phylogenetic analysis of 47 chloroplast genomes clarifies the contribution of wild species to the domesticated apple maternal line', *Molecular Biology and Evolution* 30(8) (2013), 1751–60

Papandreou, Vasiliki et al., 'Volatiles with antimicrobial activity from the roots of Greek *Paeonia* taxa', *Journal of Ethnopharmacology* 81(1) (2002), 101–04

Pavord, Anna, *The Tulip* (Bloomsbury, London, 2004)

Potter, Jennifer, *The Rose* (Atlantic Books, London, 2012)

Ramírez, Santiago R. et al., 'Dating the origin of the Orchidaceae from a fossil orchid with its pollinator', *Nature* 448 (2007), 1042–45

Robinson, Benedict S., 'Green seraglios: tulips, turbans, and the global market', *Journal for Early Modern Cultural Studies* 9(1) (2009), 93–122

Sanders, Rosanne and Harry Baker, *The Apple Book* (Frances Lincoln, London, 2010)

Sanford, Martin, *The Orchids of Suffolk: An Atlas and History* (Suffolk Naturalists' Society, 1991)

Shen-Miller, J., 'Sacred Lotus, the long-living fruits of China Antique', *Seed Science Research* 12 (2002), 131–43

Shephard, Sue, *Seeds of Fortune: A Gardening Dynasty* (Bloomsbury, London, 2003)

Tengberg, M., 'Beginnings and early history of date palm garden cultivation in the Middle East', *Journal of Arid Environments* 86 (2012), 139–47

Terral, Jean Frédéric et al., 'Insights into the historical biogeography of the date palm (*Phoenix dactylifera* L.) using geometric morphometry of modern and ancient seeds', *Journal of Biogeography* 39(5) (2012), 929–41

Thompson, Earl A., 'The tulipmania: fact or artifact?', *Public Choice* 130(1) (2007), 99–114

Ward, Cheryl, 'Pomegranates in eastern Mediterranean contexts during the Late Bronze Age', *World Archaeology* 34(3) (2003), 529–41

Widrlechner, Mark P., 'History and utilization of *Rosa damascena*', *Economic Botany* 35(1) (1981), 42–58

Wonders of Nature

Barthlott, Wilhelm et al., *The Curious World of Carnivorous Plants* (Timber Press, Portland, 2008)

Baum, David A. et al., 'Biogeography and floral evolution of baobabs (*Adansonia*, Bombacaceae) as inferred from multiple data sets', *Systematic Biology* 47(2) (1998), 181–207

Blackman, Benjamin K. et al., 'Sunflower domestication alleles support single domestication center in eastern North America', *Proceedings of the National Academy of Sciences* 108(34) (2011), 14,360–65

Colquhoun, Kate, *A Thing in Disguise: The Visionary Life of Joseph Paxton* (Harper Perennial, London, 2009)

Crane, Peter, *Ginkgo* (Yale University Press, New Haven & London, 2013)

Davis, Charles C. et al., 'The evolution of floral gigantism', *Current Opinion in Plant Biology* 11(1) (2008), 49–57

Dilcher, David L. et al., 'Welwitschiaceae from the Lower Cretaceous of northeastern Brazil', *American Journal of Botany* 92(8) (2005), 1294–310

Ervik, F. and Jette T. Knudsen, 'Water lilies and scarabs: faithful partners for 100 million years?', *Biological Journal of the Linnean Society* (2003), 539–43

Gandolfa, M. A., et al., 'Cretaceous flowers of Nymphaeaceae and implications for complex insect entrapment pollination mechanisms in early Angiosperms', *Proceedings of the National Academy of Sciences* 101(21) (2004), 8056–60

Henschel, Joh R. and Mary K. Seely, 'Long-term growth patterns of *Welwitschia mirabilis*, a long-lived plant of the Namib Desert', *Plant Ecology* 150 (2000), 7–26

Hepper, F. Nigel, *Pharaoh's Flowers: The Botanical Treasures of Tutankhamun* (HMSO, London, 1990)

Holway, Tatiana, *The Flower of Empire: The Amazon's Largest Water Lily, the Quest to make it Bloom, and the World It Helped Create* (Oxford University Press, Oxford & New York, 2013)

Huxley, Anthony, *Green Inheritance: Saving the Plants of the World* (University of California Press, Berkeley, 2005)

Jaarsveld, Ernst van and Uschi Pond, *Uncrowned Monarch of Namib (Kronenlose Herrscherin der Namib:* Welwitschia mirabilis*)*, (Penrock Publishers, Cape Town, 2013)

Li, C. et al, 'Direct sun-driven artificial heliotropism for solar energy harvesting based on photo-thermomechanical liquid crystal elastomer nanocomposite', *Advanced Functional Materials* 22(24) (2012), 5166–74

Lloyd, Francis, *The Carnivorous Plants* (Dover Publications, New York, 1976)

Pakenham, Thomas, *The Remarkable Baobab* (Weidenfeld & Nicolson, London, 2004)

Seymour, Roger S. and Philip G. D. Matthews, 'The role of thermogenesis in the pollination biology of the Amazon waterlily *Victoria amazonica*', *Annals of Botany* 98(6) (2006), 1129–35

Smith, Bruce D., 'The cultural context of plant domestication in eastern North America', *Current Anthropology* 52 S4 (2011), 471–84

Swinscow, T. D. V., 'Friedrich Welwitsch, 1806–72, a centennial memoir', *Biological Journal of the Linnean Society* 4(4) (1972), 269–89

Wickens, G. E., *The Baobab: Africa's Upside-Down Tree*, (Royal Botanic Gardens, Kew, University of Chicago Press, Chicago, 1982)

Sources of Quotations

p. 28 Nina V. Fedoroff, 'Prehistoric GM corn', *Science*, 302 (2003), 1158–59; p. 34 W. H. McNeill, 'How the potato changed the world's history', *Social Research*, 66(1) (1999), 67–83; p. 41 Lindsey Williams, *Neo Soul* (Penguin, New York, 2006), quoted in Ken Albala, *Beans: A History* (Berg, New York, 2007), 125; p. 44 Alfred Russel Wallace, *The Malay Archipelago* (Macmillan & Co., London, 1869), 233; p. 45 Pliny, *Natural History*, 18.43, trans. J. Bostock and H. T. Riley (Bohn, London, 1856); p. 49 Columella, *De Re Rustica*, V, 8; p. 52 Diogenes Laertius, *Lives of Eminent Philosophers*, 1.8, Anacharsis, ed. R. D. Hicks (Harvard University Press, Cambridge, MA, 1925); p. 58 John Gerard, *Herball* (London, 1636), 152; p. 61 John Milton, Paradise Lost II, 639–40; p. 73 Lewis Carroll, *Through the Looking-Glass and What Alice Found There* (Macmillan & Co., London, 1872), 75; p. 76 Pliny, *Natural History*, 19.42, trans. J. Bostock and H. T. Riley (Bohn, London, 1856); p. 78 John Gerard, *Herball* (London, 1636), 884; p. 80 Elizabeth David, *An Omelette and a Glass of Wine* (Penguin, London, 1984); p. 85 George Young, *A Treatise on Opium: Founded Upon Practical Observations* (London, 1753), 77; p. 89 Thomas Sydenham, *On Epidemics (Epistolae responsoriae) (Letters & Replies)*; p. 92 Jurg A. Schneider, in Robert Woodson, Jr., et al., *Rauwolfia: Botany, Pharmacognosy, Chemistry, & Pharmacology* (Little, Brown, Boston, 1955); p. 98 J. D. Hooker, *Illustrations of Himalayan Plants* (Lovell Reeve, London, 1855), Plate XIX; p. 103 quoted in Pierre Laszlo, *Citrus, A History* (University of Chicago Press, Chicago, 2007), 7; p. 108 Margaret Sanger, 'A Parents' Problem or a Woman's?', *The Birth Control Review* (1919), 6; p. 110 Philip Miller, *The Gardener's Dictionary* (London, 1768); p. 125 David Christy, *Cotton is king: or, The culture of cotton, and its relation to agriculture, manufactures and commerce* (Moore, Wilstach, Keys & Co., Cincinnati, 1855); p. 128 Peter Osbeck, *A Voyage to China and the East Indies* (London, 1771), 270; p. 130 Thomas Sheraton, *The Cabinet Dictionary*, 1803; p. 134 Lu Tung, quoted in Roy Moxham, *Tea: Addiction, Exploitation and Empire* (Constable, London, and Carroll & Graf, New York, 2003), 56; p. 138 Jonathan Swift, letter, quoted in Mark Pendergrast, *Uncommon Grounds: The History of Coffee and how it Transformed the World* (Basic Books, New York, 2010), 3; p. 154 *Hymns of the Atharva Veda*, Ralph T. H. Griffith (Luzac and Co., London, 1895); p. 151 King James I of England, *A Counter-blaste to tobacco*, 1604; p. 154 Elijah Bemiss, *The Dyer's Companion* (Evert Duyckinck, New York, 1815), 105; p. 156 quoted in Charles C. Mann, *1493: How Europe's Discovery of the Americas Revolutionized Trade, Ecology and Life on Earth / Uncovering the New World Columbus Created* (Granta, London, and Knopf, New York, 2011), 308; p. 160 Pliny, *Natural History*, 12.12, trans. J. Bostock and H. T. Riley (Bohn, London, 1856); p. 166 Henry David Thoreau, *The Maine Woods* (Tickner and Fields, Boston, 1864), 231; p. 168 John Steinbeck, *Travels with Charley: In Search of America* (Viking Books, New York, 1962); p. 174 Marion Sinclair, 'Kookaburra'; p. 176 J. D. Hooker, letter to Sir William Hooker, 1849, in L. Huxley, *Life and Letters of Sir Joseph Dalton Hooker* (John Murray, London, 1918), I, 256; p. 178 William Dampier, *A New Voyage Round the World*, Vol. 1 (London, 1697), 54; p. 183 D. T. Suzuki, *Mysticism: Christian and Buddhist* (Harper and Brothers, New York, 1957), 121; p. 188 Virgil, *Georgics*, II, trans. H.R. Fairclough (Harvard University Press, Cambridge MA, 1999); p. 198 quoted in Jennifer Potter, *The Rose* (Atlantic Books, London, 2012), 246; p. 202 quoted in Mike Dash, *Tulipomania* (Gollancz, London, and Crown Publishers, New York, 1999), 87; p. 216 Michel Adanson, *A Voyage to Senegal* (London, 1759), 96; p. 218 J. D. Hooker letter to T. H. Huxley, in L. Huxley, *Life and Letters of Sir Joseph Dalton Hooker* (John Murray, London, 1918), II, 25; p. 219 W. P. Hiern, *Catalogue of the African plants collected by Dr. Friedrich Welwitsch in 1853–61* (British Museum, London, 1896), I, xiii; p. 222 Alfred Russel Wallace, *The Malay Archipelago* (Macmillan & Co., London, 1869), 91; p. 226 John Gerard, *Herball* (London, 1636), 751; p. 228 J. W. von Goethe, 'Ginkgo biloba', 1815, from Sigfried Unseld, *Goethe and the Ginkgo: A Tree and a Poem*, trans. Kenneth J. Northcott (University of Chicago Press, Chicago & London, 2003).

Sources of Illustrations

All images are © Trustees of the Royal Botanic Gardens, Kew, unless otherwise stated.

CLA&A = Collection of the Library, Art & Archives – © Trustees of the Royal Botanic Gardens, Kew.

Half-title: CLA&A; Frontispiece: CLA&A; Title-page: K000844466 Herbarium Kew; 4a A. H. Church, *Food-grains of India* (1886), fig. 16; 4b E. Benary, *Album Benary* (1876), I, Tab. I; 5al detail, see p. 83; 5ar detail, see p. 102; 5b detail, see p. 111; 6al detail, see p. 147; 6ar detail, see p. 131; 6bl detail, see p. 140; 6br detail, see p. 179; 7a CLA&A; 7bl detail, see p. 189; 7br John Day Scrapbooks, CLA&A; 9 Roxburgh Collection, CLA&A; 10r & l CLA&A; 11 H. van Reede tot Drakestein, *Hortus Malabaricus* (1678) Pars. I, Tab. 37; 12a J.-J. Grandville, *Les fleurs animées* (1847) vol. I; 12b P.-J. Buc'hoz, *Collection precieuse et enluminée des fleurs* (1776), vol. 2, pl. XII; 13 Roxburgh Collection, CLA&A; 14, 15l CLA&A; 15r A. Targioni Tozzetti, *Raccolta di fiori frutti ed agrumi* (1825), Pl. 22; 16 J. Rea, *A Complete Florilege* (1665), frontispiece; 17 E. Benary, *Album Benary* (1879), VI, Tab. XXIII; 18l J. Metzger, *Europaeische Cerealien* (1824), Tab. VI; 18r Vilmorin-Andrieux and cie, *Les meilleurs blés* (1880), p. 135; 19, 20 CLA&A; 23 J. J. Plenck, *Icones Plantarum Medicinalium* (1794) Centuria VI, Tab. 557; 24 CLA&A; 25l J. Metzger, *Europaeische Cerealien* (1824), Tab. XIX; 25r CLA&A; 26 A. H. Church, *Food-grains of India* (1886), fig. 28; 28 A. F. Frézier, *A voyage to the South-sea, and along the coasts of Chili and Peru* (1717), pl. 10; 29 F. G. Hayne, *Getreue Darstellung und Beschreibung der in der Arzneykunde gebräuchlichen Gewächse* (1830), Vol. 11, pl. 45; 30l J. J. Plenck, *Icones Plantarum Medicinalium* (1803), Centuria VII, Tab. 694; 30r E. Benary, *Album Benary* (1876), IV, Tab. XV; 31 CLA&A; 33 K000478459 Herbarium Kew; 35 CLA&A; 36 N. F. Regnault, *La Botanique* (1774), Tome I, pl. 33; 37 W. G. Mortimer, *Peru: History of Coca* (1901), p. 196; 38 M. E. Descourtilz, *Flore pittoresque et médicale des Antilles* (1829) Tome VIII, pl. 545; 39 *Curtis's Botanical Magazine* (1838–39), vol. 65 (new ser., v. 12), Tab. 3641; 40 CLA&A; 42 H. van Reede tot Drakestein, *Hortus Malabaricus* (1688) Pars. 8, Tab. 41; 43 CLA&A; 44 J. J. Plenck, *Icones Plantarum Medicinalium* (1803), Centuria VII, Tab. 656; 45 H. van Reede tot Drakestein, *Hortus Malabaricus* (1692) Pars. 11, Tab. 22; 46 P. de' Crescenzi, *De omnibus agriculturae partibus* (1548) Liber II, p. 43; 47l W. Harte, *Essays on Husbandry* (1770, 2nd ed.) pl. V; 47r J. Metzger, *Europaeische Cerealien* (1824), Tab. XII; 48 CLA&A; 51 P. d'Aygalliers *L'olivier et l'huile d'olive* (1900), p. 257; 52 P. de' Crescenzi, *De omnibus agriculturae partibus* (1548) Liber IIII, p. 117; 53 P.-J. Redouté, *Choix des plus belles fleurs* (1827–33), pl. 24; 55 J. P. de Tournefort, *A Voyage into the Levant* (1741), p. 396; 56 E. Benary, *Album Benary* (1879), VI, Tab. XXIV; 57l CLA&A; 57r N. F. Regnault, *La Botanique* (1774), Tome 2, pl. 91; 58 Economic Botany Collection, Kew, EBC 78362; 59 F. E. Köhler, *Köhler's Medizinal-Pflanzen* (1887), Band II, Tab. 164; 60, 61 CLA&A; 62, 63 G. T. Burnett, *Medical Botany* (1835 new ed.), Vol. II, Pl. 104 and Pl. 95; 64 H. van Reede tot Drakestein, *Hortus Malabaricus* (1688) Pars. 7, Tab. 12; 65a Economic Botany Collection, Kew, EBC 78869; 65b Marianne North, 119. *Foliage, Flowers and Fruit of the Nutmeg Tree, and Humming Bird, Jamaica*; 66 CLA&A; 67 E. Benary, *Album Benary* (1877), V, Tab. XVII; 69 E. Benary, *Album Benary* (1879), VI, Tab. XXII; 70l&r J. J. Plenck, *Icones Plantarum Medicinalium*, (1789), Centuria III, Tab. 253 and Tab. 254; 72 E. Benary, *Album Benary* (1879), VI, Tab. XXI; 73 CLA&A; 74 E. Benary, *Album Benary* (1876), I, Tab. I; 75l K000914166 Herbarium Kew; 75r Agricultural Society of Japan, *The Useful Plants of Japan* (1895), Chap. XIV, No. 308; 76 J. Gerard, *Herball* (1633), Lib. 2, Chap. 457, p. 1110; 77 E. Blackwell, *A curious herbal* (1739), Vol. II, pl. 332; 79 J. J. Plenck, *Icones Plantarum Medicinalium* (1812), Centuria VIII, fasc. 2, Tab. 707; 80 *Curtis's Botanical Magazine* (1828), vol. 55 (new ser., v. 2), Tab. 2814; 81 E. Benary, *Album Benary* (1879), VI, Tab. XXIV; 82 Economic Botany Collection, Kew, EBC 41265; 83l Roxburgh Collection, CLA&A; 83r G. W. Knorr, *Thesaurus rei herbariæ hortensisque universalis* (1770), Tome I, Pars. 2, Tab. A14; 84 CLA&A; 85 J.-J. Grandville, *Les fleurs animées* (1847) vol. I; 86 'Poppy Seed Head', © Brigid Edwards, part of the Shirley Sherwood Collection; 87 J. Stephenson, *Medical Botany* (1834–36), Vol. 3, pl. 159; 88 J. J. Plenck, *Icones Plantarum Medicinalium* (1789), Centuria II, Tab. 131; 89 Economic Botany Collection, Kew, EBC 52445; 91 Tanaka Yoshio & Ono Motoyoshi, *Somoku-Dzusetsu; or, an iconography of plants indigenous to, cultivated in, or introduced into Nippon (Japan)* (1874), no. 26; 92 CLA&A; 93 H. van Reede tot Drakestein, *Hortus Malabaricus* (1686) Pars. 6, Tab. 47; 94 W. G. Mortimer, *Peru: History of Coca* (1901), p. 89; 95 K000700870 Herbarium Kew; 96 J. J. Plenck, *Icones Plantarum Medicinalium* (1789), Centuria II, Tab. 117; 97 Economic Botany Collection, Kew, EBC 49120; 98 A. Kircher, *China Illustrata* (1667), p. 184; 99 CLA&A; 101 J. J. Plenck, *Icones Plantarum Medicinalium* (1812), Centuria VIII, Tab. 701; 102 Roxburgh Collection, CLA&A; 104l J. C. Volkamer, *Nürnbergische Hesperides* (1708–14), Vol. I, p. 164 a; 104r T. Moore, *The Florist and Pomologist* (1877), f. p. 205; 106 L. Fuchs, *De Historia Stirpium* (1551) Cap XLIX, p. 143; 107 *Curtis's Botanical Magazine* (1818), vol. 45, Tab. 1975; 109 M. Scheidweiler, *L'Horticulteur Belge* (1837), No. 76; 111 P.-J. Redouté, *Choix des plus belles fleurs* (1827–33), pl. 41; 112 Roxburgh Collection, CLA&A; 113l CLA&A; 113r P. Sonnerat, *Voyage aux Indes orientales et à la Chine* (1782), Tome. 1, pl. 26; 114 CLA&A; 115 F. Antoine, *Die Coniferen* (1841) Vol. V, Tab. XXIII; 116 F. A. Michaux, *The North American Sylva* (1865), Vol. I, Pl. 2; 117 CLA&A; 118 T. Nuttall, *The North American Sylva* (1865), Vol. IV, ii; 119 British Library Public Domain Image, Harley 4425, f. 22, http://molcat1.bl.uk/IllImages/BLCD/mid/c133/c13324-62.jpg; 121 N. F. Regnault, *La Botanique* (1774) Tome II, Pl. 79; 123 J. J. Plenck, *Icones Plantarum Medicinalium* (1812) Centuria VIII, Tab. 706; 124 Roxburgh Collection, CLA&A; 126 Economic Botany Collection, Kew, EBC 73588; 127 Roxburgh Collection, CLA&A; 128 Economic Botany Collection, Kew, EBC 67854; 129 E. M. Satow, *The Cultivation of Bamboos in Japan* (1899); 130 Economic Botany Collection, Kew, EBC 73892; 131 F. G. Hayne, *Getreue Darstellung und Beschreibung der in der Arzneykunde gebräuchlichen Gewächse* (1805), Vol. 1, pl. 19; 132 CLA&A; 133l J. Cowell, *The Curious and Profitable Gardener* (1730), foldout plate; 133r CLA&A; 134 Economic Botany Collection, Kew, EBC 66450; 135 CLA&A; 136l J. C. Volkamer, *Nürnbergische Hesperides* (1708–14), Vol. II, p. 145; 136r S. Ball, *An Account of the Cultivation and Manufacture of Tea in China* (1848), Frontispiece; 137 CLA&A; 138 P. S. Dufour, *Traitez nouveaux & curieux du café, du thé et du chocolate* (1688), p. 15; 139 F. G. Hayne, *Getreue Darstellung und Beschreibung der in der Arzneykunde gebräuchlichen Gewächse* (1825), Vol. 9, Pl. 32; 140 CLA&A; 141 F. Thurber, *Coffee: from Plantation to Cup* (1881); 142 CLA&A; 143 Economic Botany Collection, Kew, EBC 40591; 145 A. H. Church, *Food-grains of India* (1886), fig. 14; 147 B. Hoola van Nooten, *Fleurs Fruits et Feuillages choisis de la flore et de la pomone de l'île de Java* (1863); 148 Economic Botany Collection, Kew, EBC 56321; 150 J. J. Plenck, *Icones Plantarum Medicinalium* (1788) Centuria I, Tab. 99; 151 P. Sonnerat, *Voyage aux Indes orientales et à la Chine* (1782), vol. 1, pl. 8; 153 B. Stella, *Il Tabacco Opera* (1669), p. 204; 154 J. G. Stedman, *Narrative of a five years' expedition against the revolted Negroes of Surinam* (2nd ed., 1806), Vol. II; 155 N. F. Regnault, *La Botanique* (177) Tome II, pl. 47; 157 F. E. Köhler, *Köhler's Medizinal-Pflanzen* (1887), Band III, pl. 8; 158 Economic Botany Collection, Kew, EBC 44097; 160 L. Colla, *Memoria sul Genere Musa e Monografia de Medesimo* (1820), end plate; 161 B. Hoola van Nooten, *Fleurs Fruits et Feuillages choisis de la flore et de la pomone de l'île de Java* (1863); 163 F. E. Köhler, *Köhler's Medizinal-Pflanzen* (1887), Band III, pl. 77; 164 CLA&A; 165l *Curtis's Botanical Magazine* (1859), vol. 85 (ser. 3, v. 15), Tab. 5146; 165r Marianne North, 563. *A Mangrove Swamp in Sarawak, Borneo*; 167 G. T. Burnett, *Medical Botany* (1835) Vol. II, Pl. LXXV (2); 169 Marianne North, 173. *Under the Redwood Trees at Goerneville, California*; 170 Marianne North, 185. *Vegetation of the Desert of Arizona*; 171 *Curtis's Botanical Magazine* (1892), vol. 118 (ser. 3, v. 4), Tab. 7222; 173 *Voyage de la corvette L'astrolabe exécuté pendant les années 1826–1829* (1833) Atlas, Pl. 10; 174 CLA&A; 175 Conrad Loddiges & Sons, *The Botanical Cabinet* (1821), Vol. 6 Tab. 501; 177l J. D. Hooker, *Himalayan Journals* (1854), Vol. II, Pl. VII (Frontispiece); 177r J. D. Hooker, *The Rhododendrons of Sikkim-Himalaya* (1849), Tab. 1; 179 J. J. Plenck, *Icones Plantarum Medicinalium* (1789) Centuria II, Tab. 359; 180 CLA&A; 181l J. J. Plenck, *Icones Plantarum Medicinalium* (1789) Centuria II, Tab. 376; 181r A. Targioni Tozzetti, *Raccolta di fiori frutti ed agrumi* (1825), Pl. 5; 182 CLA&A; 183 A. Kircher, *China Illustrata* (1667), p. 140; 184 Economic Botany Collection, Kew, EBC 41216; 185 R. Duppa, *Illustrations of the Lotus of the Ancients, and Tamara of India* (1816); 186 E. Kaempfer, *Amoenitatum Exoticarum* (1712), Fasc. IV, p. 747; 187 J. J. Plenck, *Icones Plantarum Medicinalium* (1812), Centuria VIII, Tab. 726; 188 E. Kaempfer *Amoenitatum Exoticarum* (1712), Fasc. III, p. 641; 189 F. E. Köhler, *Köhler's Medizinal-Pflanzen* (1890), Band II, Tab. 175; 190 A. Poiteau, *Pomologie française* (1846), Tome II, Grenadier Pl. 1; 191 Maria Sibylla Merian, *Metamorphosis insectorum Surinamensium* (1705), Pl. 9; 192 J. H. Knoop, *Pomologie, ou Description des meilleures sortes de pommes et de poires* (1771), Tab. V; 193 M. Bussato, *Giardino di agricoltura* (1592), Cap. 30; 195 J. J. Plenck, *Icones Plantarum Medicinalium* (1791), Centuria IV, Tab. 394; 196 Agricultural Society of Japan, *The Useful Plants of Japan* (1895), Chap. IX, No. 179; 197 CLA&A; 198 J. Gerard, *Herball* (1633), Lib. 3, p. 1261; 199 P.-J. Redouté, *Les Roses*, (1817), Vol. 1, Pl. 107; 200l CLA&A; 200r K000844514 Herbarium Kew; 202 K000844460 Herbarium Kew; 203 *Tulipa greigii* 1973, © Estate of Mary Grierson; p. 204 Robert Thornton's *Temple of Flora, or Garden of Nature* (1799–1807), Pl. 10; 206 E. Kaempfer *Amoenitatum Exoticarum* (1712), Fasc. V, p. 844; 207 CLA&A; 208 John Day Scrapbooks, volume 44, p. 13; 209 CLA&A; 210 F. Bauer, *Illustrations of Orchidaceous Plants* (1830–38) Part 2, Tab. 11; 212 P.-J. Redouté, *Choix des plus belles fleurs* (1827–33), Pl. 22; 213, 214 CLA&A; 215l Roxburgh Collection, CLA&A; 215r *Curtis's Botanical Magazine* (1863), vol. 89 (ser. 3, v. 19), Tab. 5369; 216 *Curtis's Botanical Magazine* (1828), vol. 55 (new ser., v. 2), Tab. 2791; 217 Thomas Baines Collection, Royal Botanic Gardens, Kew; 219 *Curtis's Botanical Magazine* (1863), vol. 89 (ser. 3, v. 19), Tab. 5368; 221 *Curtis's Botanical Magazine* (1847), vol. 73 (ser. 3, v. 3), Tab. 4276; 223 *Curtis's Botanical Magazine* (1828), vol. 55 (new ser., v. 2), Tab. 2798; 224, 225 CLA&A; 226 C. Duret, *Histoire admirable des plantes et herbes esmerueillables & miraculeuses en nature* (1605), p. 253; 227 CLA&A; 228 E. Kaempfer, *Amoenitatum Exoticarum* (1712), Fasc. V, p. 811; 229 CLA&A.

Acknowledgments

Many thanks to our editor, Colin Ridler, at Thames & Hudson and to Gina Fullerlove, Head of Publishing at the Royal Botanic Gardens, Kew, for sharing our excitement and taking this book forward. Kew has very fine human resources as well as the most wonderful collections of art and literature devoted to plants. Julia Buckley from the Illustrations Team of the Library, Art and Archives at Kew insisted that it was a 'team' effort, but her patient, unflagging and enthusiastic help has been invaluable. Thanks to Dr Shirley Sherwood, Brigid Edwards and the estate of the late Mary Grierson for allowing us to use images.

All the Kew Library reading room staff coped with our huge request lists with a smile and provided a smooth, efficient and conversant service. UCL and Suffolk Libraries also helped made this book possible. The immensely knowledgeable Mark Nesbitt, Curator of the Economic Botany Collection, Kew, kindly read the entire text and made wonderfully piquant comments for which we are very grateful. Additional help at Kew came from Christine Beard, Anna Trias Blasi, Lorna Cahill, Paul Little, Trishya Long, Barbara Lowry, Christopher Mills, Lynn Parker, Martyn Rix, Kiri Ross-Jones, Georgina Smith, Maria Vorontsova, Paul Wilkin and Lydia White. Dorian Q. Fuller and Michael D. Coe made helpful comments on parts of the text. Lin Zhang helped with Chinese poetry and translations. Any errors that remain are our own.

At Thames & Hudson Niki Medlikova's and Lauren Necati's design skills are evident, and thanks also to Sarah Vernon-Hunt and Rachel Heley.

Andy and Nancy Scull have an amazing garden in San Diego and took us to Muir Woods; Sarah Duignan asks the right kind of pointed questions and Emma Kelly just 'gets it'. Thanks to all.

Half-title: Waterlily (*Nymphaeum candida*).
Frontispiece: Himalayan rhubarb (*Rheum australe*).
Title page: Dried and pressed tulip (*Tulipa armena* var. *lycica*).
Page 4: *above* Head of 'broomcorn' (*Sorghum bicolor*); *below* cabbages (*Brassica oleracea*).
Page 5: *above left* Strychnine (*Strychnos nux-vomica*); *above right* sour lime (*Citrus aurantifolia*); *below* Madagascar periwinkle (*Catharanthus roseus*).
Page 6: *above left* Chocolate (*Theobroma cacao*); *above right* mahogany (*Swietenia mahagoni*); *below left* coffee (*Coffea arabica*); *below right* red mangrove (*Rhizophora mangle*).
Page 7: *above* Orchid (*Vanda bicolor* Griff.); *centre* frankincense (*Boswellia sacra*); *below right* orchid (*Odontoglossum triumphans*).

For Jonathan and Julie
'Vegetable love'

Helen Bynum is a historian of science and medicine. William F. Bynum is professor emeritus of the UCL Centre for the History of Medicine in London. Together they have edited *Great Discoveries in Medicine* and *Dictionary of Medical Biography*.

The University of Chicago Press, Chicago 60637
The University of Chicago Press, Ltd., London
Text and Layout © 2014 Thames & Hudson Ltd, London
Illustrations © 2014 the Board of Trustees of the Royal Botanic Gardens, Kew, unless otherwise stated
All rights reserved. Published 2014.
Printed in China

23 22 21 20 19 18 17 16 15 14   1 2 3 4 5

ISBN-13: 978-0-226-20474-1 (cloth)
ISBN-13: 978-0-226-20488-8 (e-book)
DOI: 10.7208/chicago/9780226204888.001.0001

Library of Congress Cataloging-in-Publication Data

Bynum, Helen, author.
  Remarkable plants that shape our world / Helen Bynum and William Bynum.
    pages ; cm
  ISBN 978-0-226-20474-1 (cloth : alk. paper) — ISBN 978-0-226-20488-8 (e-book) 1. Gardening. 2. Plants.
I. Bynum, W. F. (William F.), 1943– author. II. Title.
  SB450.97.B96 2014
  635—dc23
        2014010717